Navigation Channel Sedimentation Solutions

Other Titles of Interest

Planning and Design Guidelines for Small Craft Harbors, Third Edition, Task Committee on Marinas 2020 (ASCE/COPRI 2012). MOP 50 provides new, state-of-the-art guidelines for the planning, design, and development of small craft harbors. (ISBN 978-0-7844-1198-8)

Ship Channel Design and Operation, edited by Bruce L. McCartney, Laurie L. Ebner, Lyndell Z. Hales, and Eric E. Nelson (ASCE/COPRI 2005). MOP 107 provides an overview of the design process and operation of deep-draft navigation projects. (ISBN 978-0-7844-0770-7)

Inland Navigation: Locks, Dams, and Channels, edited by Bruce L. McCartney, John George, B. K. Lee, Mark Lindgren, and Frank Neilson (ASCE/COPRI 1998). MOP 94 provides information on planning, design, construction, and operation of the U.S. waterways used by barge traffic. (ISBN 978-0-7844-7033-6)

Inland Navigation: Environmental Sustainability, Task Committee on Inland Navigation of the Waterways Committee; edited by Bruce L. McCartney (ASCE/COPRI 2019). This report provides an overview of the ecosystem sustainability procedures currently used for inland waterways in the United States. (ISBN 978-0-7844-8151-6)

Navigation Engineering Practice and Ethical Standards, edited by William H. McAnally (ASCE/COPRI 2008). MOP 116 presents engineering criteria and practices for the design, operation, and management of navigation projects and shows how to integrate them with engineering ethics. (ISBN 978-0-7844-0992-3)

Inland Navigation: Channel Training Works, edited by Thomas J. Pokrefke (ASCE/COPRI 2013). MOP 124 presents design guidance on structures that reshape a river channel to create reliable depths and widths for safe and dependable vessel transit. (ISBN 978-0-7844-1253-4)

ASCE Manuals and Reports on Engineering Practice No. 156

Navigation Channel Sedimentation Solutions

Sponsored by
Navigation Channel Sedimentation Task Committee of the
Navigation Engineering Subcommittee of the
Waterways Committee of the
Coasts, Oceans, Ports, & Rivers Institute of the
American Society of Civil Engineers

Published by the American Society of Civil Engineers

Library of Congress Cataloging-in-Publication Data

LCCN 2023036171
Cataloging-in-Publication Data on file with the Library of Congress

Published by American Society of Civil Engineers
1801 Alexander Bell Drive
Reston, Virginia 20191-4382
www.asce.org/bookstore | ascelibrary.org

***Errata*: Errata, if any, can be found at https://doi.org/10.1061/9780784485149.**

ISBN 978-0-7844-1616-7 (print)
ISBN 978-0-7844-8514-9 (PDF)
ISBN 978-0-7844-8515-6 (ePub)

Manufactured in the United States of America.
27 26 25 24 23 1 2 3 4 5

MANUALS AND REPORTS ON ENGINEERING PRACTICE

(As developed by the ASCE Technical Procedures Committee, July 1930, and revised March 1935, February 1962, and April 1982)

A manual or report in this series consists of an orderly presentation of facts on a particular subject, supplemented by an analysis of limitations and applications of these facts. It contains information useful to the average engineer in his or her everyday work, rather than findings that may be useful only occasionally or rarely. It is not in any sense a "standard," however, nor is it so elementary or so conclusive as to provide a "rule of thumb" for nonengineers.

Furthermore, material in this series, in distinction from a paper (which expresses only one person's observations or opinions), is the work of a committee or group selected to assemble and express information on a specific topic. As often as practicable, the committee is under the direction of one or more of the Technical Divisions and Councils, and the product evolved has been subjected to review by the Executive Committee of the Division or Council. As a step in the process of this review, proposed manuscripts are often brought before the members of the Technical Divisions and Councils for comment, which may serve as the basis for improvement. When published, each manual shows the names of the committees by which it was compiled and indicates clearly the several processes through which it has passed in review, so that its merit may be definitely understood.

In February 1962 (and revised in April 1982), the Board of Direction voted to establish a series titled, "Manuals and Reports on Engineering Practice" to include the manuals published and authorized to date, future Manuals of Professional Practice, and Reports on Engineering Practice. All such manual or report material of the Society would have been refereed in a manner approved by the Board Committee on Publications and would be bound, with applicable discussion, in books similar to past manuals. Numbering would be consecutive and would be a continuation of present manual numbers. In some cases of joint committee reports, bypassing of journal publications may be authorized.

A list of available Manuals of Practice can be found at https://ascelibrary.org/page/books/s-mop.

DEDICATION

This manual is dedicated to Bruce L. McCartney, P.E., BC.NE, M.ASCE. Bruce has been a major force in the development of modern navigation engineering practice throughout his careers in the US Coast Guard and US Army Corps of Engineers and decades of active ASCE/COPRI Waterways Committee and Navigation Engineering Subcommittee membership. He wrote and guided half a dozen ASCE manuals on navigation engineering and contributed to scores of conferences and journal papers. His advocacy led to the discipline being recognized as a Board Certified Engineering Specialty by the Academy of Coastal, Ocean, Port, and Navigation Engineers. ASCE recognized Mr. McCartney's accomplishments with its prestigious Hans Einstein Award in 2013. His colleagues in navigation engineering cherish him as a friend, mentor, and continuing source of knowledge and inspiring ideas.

CONTENTS

NAVIGATION CHANNEL SEDIMENTATION TASK COMMITTEE

The Navigation Channel Sedimentation Task Committee of the Waterways Committee of ASCE's Coasts, Oceans, Ports and Rivers Institute prepared this manual. Task Committee Member authors are

Tanya M. Beck, Ph.D.
Coastal and Hydraulics Laboratory
Engineer Research and Development Center
US Army Corps of Engineers

Lynn Bocamazo, P.E., BC.CE
New York District (Retired)
US Army Corps of Engineers

Gary L. Brown
Coastal and Hydraulics Laboratory
Engineer Research and Development Center
U.S. Army Corps of Engineers

Kristina Cydzik, P.E., LEED AP
Engineering Systems Inc.

Craig Jones, Ph.D.
Integral Consulting, Inc.

Dennis Lambert M.Eng, P.E., BC.NE, BC.PE, BC.WRE, F.ASCE
IEM

Rooni Mathew, Ph.D., P.E.
CDM Smith

William H. McAnally Ph.D., P.E., BC.NE, BC.CE, F.ASCE, ***Co-chair***
Dynamic Solutions, LLC
and Geosystems Research Institute (Emeritus)
Mississippi State University

Nancy J. Powell, P.E., BC.WRE, BC.NE, M.ASCE
Arcadis U.S., Inc.

Thomas J. Pokrefke, P.E., BC.NE, LM.ASCE
Coastal and Hydraulics Laboratory (Emeritus)
Engineer Research and Development Center
US Army Corps of Engineers

Parmeshwar L. Shrestha, Ph.D., P.E., BC.WRE, BC.CE, ***Co-chair***
Engineering Systems Inc.

Christopher M. Wallen, P.G., A.M.ASCE
Dynamic Solutions, LLC

ACKNOWLEDGMENTS

We thank the following members of the Blue Ribbon Committee for their careful review of the document and excellent recommendations for improvements.

Earl J. Hayter
Environmental Laboratory
Engineer Research and Development Center
US Army Corps of Engineers

William Anthony (Tony) Thomas
Mobile Boundary Hydraulics, LLC

Jane McKee Smith
Coastal and Hydraulics Laboratory (Emeritus)
Engineer Research and Development Center
US Army Corps of Engineers

Special thanks to Carol McAnally for copyediting the draft document, to COPRI staff Tom Chase and Sean Herpolsheimer for their steadfast support to the committee, ASCE Acquisitions Editor Natalie Webster and Senior Manager Michie Gluck for shepherding this manual through publication.

EXECUTIVE SUMMARY

INTRODUCTION

This manual of practice describes navigation channel sedimentation, lists solutions to sedimentation problems in those channels, and recommends best practices for predicting navigation channel sedimentation responses—sediment deposition to, and erosion of, the channel bed plus surrounding bed and banks—resulting from those solutions.

Navigation is an integral part of the global and US economy, contributing more than $500 billion to the US gross domestic product. The United States spends more than $2 billion annually on dredging, mostly on navigation channel maintenance. Yet, without dredging or other solutions, a lack of adequate navigation channel depth impedes commerce and creates hazardous conditions for vessels. These impacts provide ample motivation to manage sedimentation processes in navigation channels.

Managing waterway sedimentation processes requires making sedimentation response predictions before new channels are constructed or existing channels are modified to reduce dredging needs. Presently available up-to-date critical review or categorization for evaluating navigation channel sedimentation solutions and prediction practices is extremely limited. This manual is intended to fill that gap through

1. Overview of channel sedimentation processes,
2. Discussion of navigation channels' classifications and typical sedimentation issues,
3. Solutions/remedies to channel sedimentation problems with sustainability concepts,
4. Channel sedimentation predictions methods
   a. Field observations
   b. Desktop analyses

   c. Physical (scale) models
   d. Numerical models
   e. Uncertainty in predictions from these methods,
5. Recommendations for best practices in using the prediction methods, and
6. Glossary of terms and acronyms.

Sedimentation

Reliable solutions to, and accurate predictions of, navigation channel sedimentation require a thorough knowledge of sediment transport processes, including the hydrologic and hydraulic forces driving those processes. Broadly speaking, the principal external forces of interest are water, wind, and gravity. In riverine locations, gravity is the dominant external energy force, and the hydrodynamic equations need only to translate that force into flow velocity and boundary shear. The computation is often uncomplicated by variations in fluid densities or wind energy. In coastal locations, wind and tides may be the dominant external energy forces of interest. However, the water properties of density and temperature may vary in the vertical and horizontal directions.

Also broadly speaking, sediment transport in rivers, lakes, estuaries, and coastal settings involves the advection and diffusion of sediment originating from freshwater and/or marine sources, and the bed-water exchange processes of erosion and deposition. Transport is a function of the physical forcings acting upon the sediment, including the fluid forces of barotropic flow (water density a function of pressure only), baroclinic circulation (water density a function of pressure, temperature, and constituent concentrations), freshwater flow, waves, and winds. In addition to impacting transport directly, some of these forcings may also have secondary effects via nonlinear interactions with the others. Anthropogenic activities, primarily vessel traffic, can also directly impact sediment transport through vessel wakes and propwash.

The sources, transport, and eventual sinks of sediment represent a complex interplay of long- and short-term processes that must be understood to predict future behavior. In general, prediction of sedimentation in navigation channels involves shorter-term processes on the order of seasonal to annual timescales that deliver sediment from source areas to the channel. Nevertheless, longer-term processes must be considered in projects designed for 50-year-life spans and most often employed for far longer periods.

Sediment can be classified as cohesive or noncohesive. Cohesive sediment behavior is affected by interparticulate attractive forces that cause them to aggregate into flocs containing hundreds to millions of individual mineral grains. Cohesive sediments are generally comprised of clay-sized (< 2 μm) and silt-sized (< 62 μm) particles. High-concentration cohesive sediment suspensions may exhibit non-Newtonian behavior when in a fluid-like state. Noncohesive sediments are sand and gravel-sized material (> 62 μm) plus coarse silts where the particles exert only weight, friction, and impact forces on one another. Overall, noncohesive sediment transport properties are much better understood than cohesive sediment properties because of the difficulty of quantitatively predicting the role of interparticulate forces in cohesive sediment.

Moving water, whether because of river flow, tides, or waves, exerts a shear stress on the sediment bed. The shear stress, if large enough, may mobilize or erode sediment from the sediment bed and suspend it into the water column. Advection moves suspended sediment according to the local water velocity, whereas turbulent diffusion spreads sediment at rates dependent on concentration gradients and turbulence. Further, sediment (both noncohesive and cohesive) may also be transported as bedload—moving in proximity to, and in frequent contact with, the bed. When sediments move into lower-energy regions where turbulence and stresses are not high enough to keep them moving, the sediment settles and deposits.

Navigation Channels

Navigation channels can be categorized by depth: shallow-draft and deep-draft channels with 14 or 15 ft of navigable depth as the most commonly used dividing line. In the United States, shallow-draft channels typically provide 9 to 12 ft of depth and are found in the inland waterway network, the Gulf and Atlantic Intracoastal Waterways, and small ports nationwide. Portions of the Gulf Intracoastal Waterway have an authorized depth of 16 ft but are still considered shallow-draft channels. Deep-draft channels offer navigable depths of 18 to 50 ft and more.

Navigation channel sedimentation problems are most often deposition or excessive erosion. Natural water depths are often too shallow to allow safe navigation and dredging them to provide sufficient depth creates an imbalance in the relationship between sediment supply and sediment transport capacity, resulting in deposition. Excessive sediment erosion becomes a problem when channel banks, nearby land, structures, or buried utilities are threatened.

The alignment and configuration (e.g., depth, slope, width) of a navigation channel affects sediment transport in several ways. In general, a deep trench channel cut into a natural alluvial plain or coastal morphology will disrupt the movement of sediment because of reduced current speed and/or wave effects in the deeper water of the channel. A channel aligned with the direction of dominant sediment transport may experience less deposition than one perpendicular to the transport direction. Thus, a channel following the thalweg of a natural waterway will tend to experience less deposition than a channel that crosses the main transport path.

Channels approaching and passing through riverine navigation locks often present sedimentation challenges, because the pooled water on the upstream side (upper pool) provides a quiet water environment suitable for deposition. The downstream side (lower pool or tailwater) of the dam may be subject to erosion. Harbors, ports, docks, terminals, and marinas plus their connecting channels, anchorages, and turning basins experience the same sedimentation issues as navigation channels plus some unique problems related to their configuration.

Sedimentation Solutions

The variety of channel environments and sedimentation processes prevents a one-size-fits-all solution to deposition and erosion problems. Solutions to sedimentation problems may take many forms, ranging from nonstructural to structural to dredging. A useful classification system consists of three categories and seven strategies:

1. Prevention
   (a) Keep sediment in place.
   (b) Keep sediment out.
   (c) Keep sediment moving.
2. Treatment
   (a) Keep sediment navigable.
   (b) Dredge and remove sediment.
   (c) Dredge and place sediment.
3. Accommodation—Adapt to sedimentation regime.

Examples within each of these categories provide an organized approach to selecting from a variety of appropriate solutions as described in Case Studies.

Prediction Methods

Channel sedimentation responses to proposed solutions can be predicted by numerous methods grouped into four general approaches: field observations, desktop analyses, physical models, and numerical models. These approaches vary in terms of data requirements, complexity, reliability, applicability, cost, and time requirements. They should be applied with full recognition of their strengths and weaknesses and the user's skill in applying them.

Field observations provide insight into the physical processes of sedimentation and data for desktop analyses, physical modeling, and numerical modeling. Basic data collected for sedimentation studies include standard time-varying morphologic, hydrodynamic, sediment, and water-quality measurements. Rates of channel deposition and erosion can be estimated by hydrographic/bathymetric surveys analysis, densiometric surveys analysis, dredging records, sediment sample dating, and test pits.

Desktop approaches represent the simplest methods for estimating sedimentation rates and can be applied relatively efficiently. Most desktop methods are empirical in nature, relying on field observations to predict future sedimentation rates. Their benefits of speed and low cost arise from assumed simplifications to physics and geometry and those same simplifications often produce lesser reliability than more rigorous modeling approaches.

Physical models are scale models constructed according to scientific scaling principles for geometric, kinematic, and dynamic similarity. They require considerable expertise, typically gained through long experience. Therefore, most sediment transport modeling is now performed using numerical modeling.

Process-based numerical methods for estimating navigation channel sedimentation rates solve differential and algebraic equations representing the physical forcings and transport processes relevant to sedimentation. Numerical sedimentation models of waterways consist of two principal parts—a modeling program that solves the appropriate equations and a digital representation of the waterway to be modeled. The digital representation of a waterway includes inputs such as a mesh that defines the physical boundaries of the waterway and initial elevations of the bed, banks, structures, and overbanks; some combination of water, sediment, and sometimes constituent inflows, outflows, and water levels at the water boundaries; bed and inflowing sediment characteristics; other pertinent driving forces; and parameters required by the computer program equations. The number of spatial dimensions—zero-dimensional (0D), one-dimensional (1D), two-dimensional (2D), or three-dimensional (3D)—constitutes a primary classification for numerical model program capabilities and applications. More spatial dimensions result in more equations to solve and that generally requires more computer time and storage space.

All quantitative predictions of natural systems, including sedimentation, have inherent uncertainty resulting from errors in the inability to know future conditions, understanding and measurement of the physical and biological processes, and approximations and simplifications of the prediction method. Engineering decisions must be made despite these uncertainties, using the best available information with a full appreciation of the uncertainty bounds on that information. Therefore, modelers are responsible for quantifying uncertainty in their results and reporting them alongside model results.

Best Practices

Channel sedimentation problems are best solved through a methodical, step-by-step approach:

1. Problem Definition—including a written set of objectives, products, and schedule with quality expectations; definition of project and process boundaries; inventory of available data; and a Conceptual Site Model of sedimentation processes.
2. Design Approach—including progressing from simplest to more rigorous solution methods, selecting appropriate tools based on objectives, time and space scales, and available resources.
3. Tools Application—using best professional practices as detailed in this manual.
4. Interpreting and Reporting—using engineering judgement, not simply relaying raw numbers, to provide interpreted findings, including uncertainty bounds and limitations that may influence decision-making.

Managing a sedimentation study includes ensuring that the aforementioned steps are followed conscientiously, that client and team communications (including written progress reports) are frequent and candid, and that a rigorous quality assurance/quality control process is followed.

CHAPTER 1
INTRODUCTION

PURPOSE

This manual describes navigation channel sedimentation, lists solutions to sedimentation problems in those channels, and recommends best practices for predicting navigation channel sedimentation responses—sediment deposition to, and erosion of, the channel bed plus surrounding bed and banks—resulting from those solutions.

SCOPE

Vanoni (1975) defines sedimentation as follows:

> Sedimentation embodies the processes of erosion, entrainment, transportation, deposition and the compaction of sediment. These are natural processes that have been active throughout geological times and have shaped the present topographic and bathymetry dimensions of our earth.

The manual focuses on sedimentation in navigation channels within rivers, canals, lakes, estuaries, and coastal waterways that allow commercial waterborne transport of people and goods and recreational use. Most, but not all, navigation channels are designated as such by government or other authorities. Ports, terminals, and anchorages located on or adjacent to these navigation channels experience the same sedimentation processes as channels. Solution procedures recommended for channels may apply to those navigation facilities as well. Overall geomorphological processes and water bodies not designated as navigation channels, such as streams, lakes, shallow coastal waters, and coastlines, are explicitly considered here only to the extent that they affect navigation channel sedimentation.

MOTIVATION

Navigation is an integral part of the global and US economy, contributing more than $500 billion to US gross domestic product (MARAD 2021). Navigation is a cost-effective way to transport many goods and can be less environmentally intrusive when compared to other transportation methods such as rail and trucks (Kruse et al. 2021). As vessels have become larger, it has become necessary to enlarge navigation channels. The United States spends more than $2 billion annually on dredging, mostly on navigation channel maintenance (USACE 2021). World-wide total dredging costs are estimated to be more than $10 billion (Verified Market Research 2022) and increasing. Yet, without dredging or other solutions, a lack of adequate navigation channel depth impedes commerce and creates hazardous conditions for vessels, resulting in financial burden on society. Sedimentation issues may impact environmental quality through increased vessel emissions (caused by longer transit times or additional trips required by light-loading vessels) and impacts on ecologically important habitats and communities. Shallow-draft channel sedimentation can decrease recreational opportunities when vessels cannot safely transit shoals. These economic, social, and environmental impacts provide ample motivation to manage sedimentation processes in navigation channels.

Charlie Little, retired USACE engineer, often said, "Complex river engineering problems often have simple, easy to understand, *wrong* answers." He arrived at this conclusion after witnessing many failed attempts to fix sedimentation problems with solutions that only made the problems worse. We have seen the same in lakes, estuaries, and coastal zones. Managing waterway sedimentation processes by trial and error often exacerbates problems, making accurate sedimentation response predictions imperative before new channels are constructed or existing channels modified. Channel modifications may include changes in dimensions and/or alignment, addition or modification of structures (e.g., dikes, jetties, breakwaters, dams), and altered flows or sediment supply (e.g., impoundments, diversions, land use and land cover change, and climate change).

An evaluation of sedimentation responses to these changes is necessary to select preferred designs and to evaluate the environmental impacts of alternative improvements. Typically, evaluation requires information on potential changes to volume, timing, and pattern of bed and bankline sediment deposition or erosion, characteristics of bottom sediment (e.g., mud, sand, gravel), suspended sediment concentrations, and sediment-associated nutrients or pollutants.

Although some guidance for evaluating navigation channel sedimentation is available in ASCE manuals of practice and US Army Corps of Engineers (USACE) engineer manuals, much of that information has not been updated in recent years and has gaps with respect to best practices, particularly with respect to predictive methods. The professional literature presents more contemporary information; however, it is scattered across a variety of publications. No presently available up-to-date critical review or categorization of navigation channel sedimentation solutions and prediction practices exists. This manual is intended to fill this gap.

PROBLEM DEFINITION

Typical channel sedimentation problems include

- Problem depositional areas requiring frequent dredging, such as Aberdeen Pool on the Tombigbee Waterway (Mississippi), Savannah Harbor (Georgia), Lower Mississippi River from Venice to the Gulf of Mexico (Louisiana), San Francisco Bay and Delta (California), and the Mouth of the Columbia River (Washington and Oregon), are regularly examined for potential sedimentation solutions. Evaluating these solutions requires predicting how much channel sediment deposition will be reduced and whether benefits to navigation exceed the costs of channel modifications (Teeter 1989, USACE 2022).
- New projects such as channel deepening to allow larger vessels, for example, those implemented along the St. Johns River (Florida) and Stockton Ship Channel (California), and flow alterations such as proposed Mississippi River diversions (Louisiana) must be evaluated for sedimentation changes as they affect channel maintenance and environmental impacts, including water quality and ecosystem services.
- Existing projects that may experience sedimentation volume and location changes as a result of dam operations, water diversions/additions, climate and land cover change, and sea-level rise. Navigation channel managers need to anticipate potential changes in channel sedimentation to predict future dredging needs, costs, and potential environmental impacts. For example, freshwater diversions may cause stratification or sediment supply changes in either the exporting or the importing waterways sufficient to substantially alter navigation channel sedimentation.

Navigation channel dredging in the United States is performed or funded primarily by USACE in the federal system of inland and coastal waterways shown in Figure 1-1. From 1963 to 2020, an average of 233 million yd^3 of material was dredged annually by USACE-owned and industry dredges to maintain safe and efficient navigation. Of the average annual amount dredged, an average of 53 million yd^3 was removed as part of new construction work. Figure 1-2 shows dredging quantities by year for fiscal years 1963 through 2020 categorized by USACE-owned and industry dredges. Costs for dredging increased over the same period from \$200,000 to about \$2.5 billion for fiscal year 2020, translating to an average 2020 unit cost of \$9.27/yd^3. Figure 1-3 shows yearly costs for 1963 to 2020. Smith (2019) reported 15 USACE studies underway to deepen navigation ports and channels to accommodate higher traffic loads and larger vessels, with seven projects under construction.

Dredge equipment availability and USACE budgets have affected the USACE's ability to maintain navigation projects to authorized depths. If navigation channels are not maintained to design dimensions and depth restrictions are imposed, shipping costs and safety are impacted. For

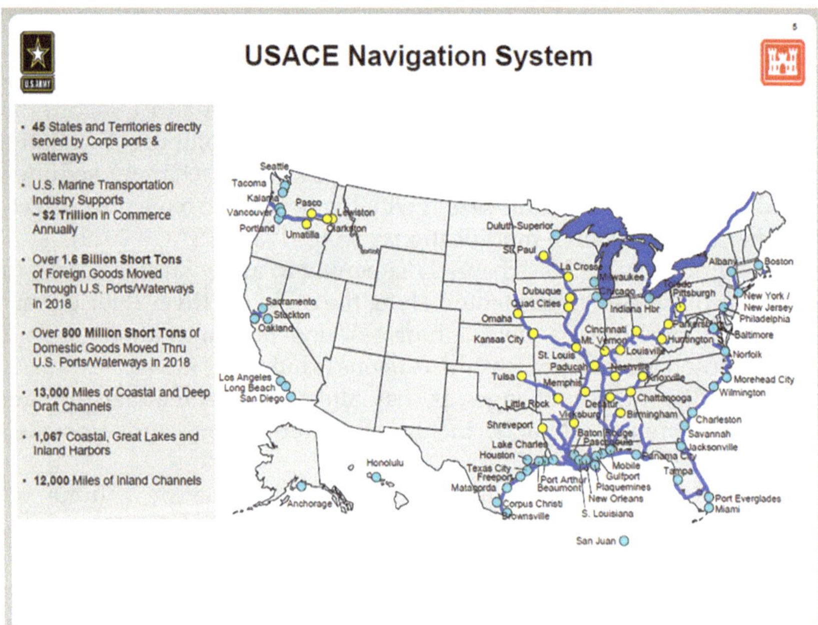

Figure 1-1. Federal navigation channels in the United States.
Source: Burroughs (2020).
Note: Blue dots indicate deep-draft ports; yellow dots indicate shallow-draft ports.

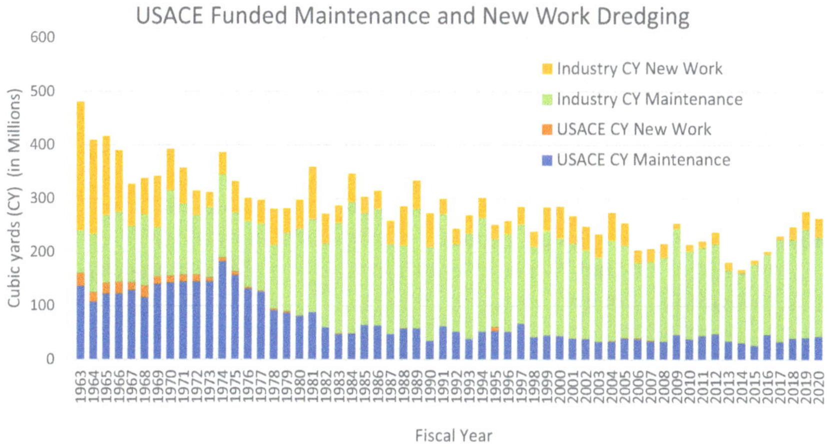

Figure 1-2. USACE-funded maintenance and new work dredging volumes. Source: USACE (2021).

example, the potential effect of draft restrictions to deep-draft vessels in Southwest Pass of the Mississippi River range from $36 million in additional annual cost for a 1 ft loss in draft to $410 million in additional annual cost for a 6 ft loss in available channel depth (Mitchell 2020). In fiscal 2016, there were channel restrictions on the Mississippi River for a total of 182 days, with a minimum estimated economic loss of $96 million (USACE 2017). An average annualized impact of $376 million is estimated for 20 USACE navigation projects disrupted by a 5 ft draft restriction (USACE 2017).

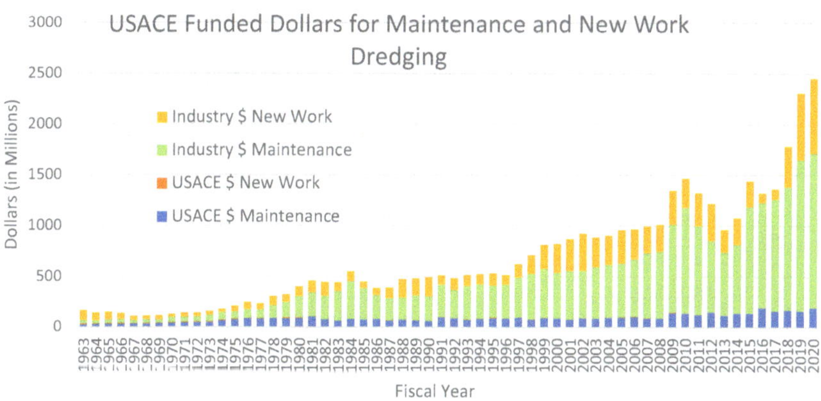

Figure 1-3. USACE-funded dollars for maintenance and new work dredging. Source: USACE (2021).

Solutions to sedimentation problems that can reduce dredging needs are available, as discussed in Chapter 4, "Solutions to Sedimentation Problems." As mentioned previously, implementing solutions on a trial-and-error basis can result in costly mistakes that worsen the problems. Appropriate predictive methods, as discussed in Chapter 5, "Prediction Methods," and properly applied as described in Chapter 6, "Best Practices," are essential to producing outcomes that allow efficient navigation, save money, and protect or even enhance environmental quality.

Methods for predicting sedimentation must be able to consider a range of significant factors, including

- Geographic, geologic, and geomorphological influences (e.g., a river flowing through steep, rocky canyons will behave differently from one flowing across a flat and broad alluvial plain);
- Hydrodynamic forces ranging from overland flow to river and lake flow to tide and wave-dominated coastal areas;
- Climatological effects such as seasonal weather patterns and changing storm frequency, intensity, and direction;
- Sediment sizes ranging from colloidal to boulders and the way in which various physical forces affect their behavior (e.g., physicochemical forces on clay mineral grains);
- Biological and chemical processes on landscapes and in waters that affect sediment behavior, including settling and erodibility; and
- Differing temporal and spatial scales among significant processes of interest (e.g., millimeters and seconds for individual sediment particles, meters and weeks for critical deposition events, and kilometers and decades for some morphological processes, watershed changes, and climate changes).

Complicating sedimentation predictions for navigation projects are practical issues, such as

- Sedimentation field data are often spatially and temporally sparse.
- Predictive methods typically produce deterministic results despite the stochastic nature of sedimentation processes, sometimes misleading decision makers.
- Both the previous points produce uncertainty in engineering estimates of sedimentation rates, which can be difficult to quantify and explain.
- Low-cost projects with inconsequential sedimentation impacts may justify only a minimum level of effort for predictions analyses. Projects with higher costs or potential for adverse consequences may justify more substantial prediction efforts. Distinguishing the appropriate level of effort for each project often confounds decision makers.

SEDIMENTATION SOLUTIONS AND PREDICTION HISTORY

Sedimentation issues in channels are as old as the practice of employing channels for commercial traffic. The port at Rhakotis, Egypt, built circa 1900 BCE became unusable because of sediment accumulation. Other ancient ports such as the Roman port of Ostia, plus ports in India, China, and East Africa, suffered similar problems, which resulted in abandonment of some of these ports. Some ports relocated to naturally deep and relatively sediment-free waters to avoid such fates.

Ancient remedies for excessive channel sedimentation advanced slowly, with early techniques using agitation dredging, such as the Chinese *hun jiang lung*, a roller with blades, drug along the bottom in waterways to disturb bed sediments. Dredging technology increased to crude dipper-and-bag to bucket machines. Training walls to scour sediment by focusing currents were constructed at the Roman Empire's port of Narbonne, France.

Historically, remedies for port sedimentation—relocation, dredging, and training structures—tended to be chosen by intuition and trial and error; however, a few solutions appear to have been based on a keen knowledge of physical processes. For example, Mehta (2018) mentions the half-tide harbor at Lothal on the Indus River, India, built around year 2350 BCE, where vessels could enter and exit only at high levels of the tide. A shallow sill limited vessel draft but also reduced the volume of near-bed sediments entering the harbor from the river. Such well-thought-out solutions were relatively rare compared with mistake-plagued solutions selected on the basis of uninformed or nonexistent estimates of future sedimentation rates.

Trial-and-error sedimentation solutions continued to be the norm until the sixteenth century, when channel designers, notably Leonardo da Vinci, began using reasoning and experimentation. Scale models of hydraulics problems were employed as early as the eighteenth century and by the nineteenth century, and mobile bed scale models came into use for sedimentation evaluation (Rouse and Ince 1957).

From the sixteenth century onward, channel designers employed rational but simplistic rules of thumb to design channels, methods which we now refer to as desktop designs. Some rules were based on channel regime theory, developed in India for application to canal sizing, but also applied to rivers and navigation channels. Other design concepts attempted to achieve specific channel velocities to achieve cross-sectional stability. James Eads employed the latter in 1875 to design jetties at the Gulf of Mexico mouth of the Mississippi River's South Pass, using his experience at other locations on the river as a guide. The jetties concentrated the river's flow to scour the channel to a depth of 30 ft, as Eads predicted (Schulz 1917).

In the twentieth century, physical (scale) models for studying sedimentation issues came into prominence, led by laboratories in Germany and then the United States. In the 1930s through 1980s, the USACE Waterways Experiment Station, in Vicksburg, Mississippi, designed and evaluated sedimentation solutions using large-scale physical models, including reproductions of Savannah River, Mississippi River Head of Passes, and Columbia River (Hudson et al. 1979, CTH 1971, Fatheree 2004).

Numerical modeling of sedimentation problems and their solutions began in the 1960s when W. Anthony Thomas of the USACE Little Rock District wrote a sedimentation analysis computer program and applied it to the Arkansas River (USACE 1995). Thomas moved to the USACE Hydrologic Engineering Center (HEC) in Davis, California, and his program became HEC-6, a one-dimensional sediment transport model of river mechanics that computed scour and deposition by simulating the interaction between flow hydraulics and the rate of sediment transport. HEC-6 was designed to be used for the analysis of long-term river and reservoir changes, including changes to navigation channel depths.

Thomas developed a research version of HEC-6 called Tabs-1 in the 1980s after he was transferred to the USACE Waterways Experiment Station. Cohesive sediment erosion algorithms were added to Tabs-1 for the Red River Navigation Study in 1986, and the rate of dredging and location of dredge discharge was added in 1991 for studies of the lower Mississippi River in the New Orleans District. Other improvements included stream network capability and additional sediment transport functions. Tabs-1 was then used to develop version 4.0 of HEC-6. The last version of HEC-6 released was version 4.1 in 1993.

In the 1990s, HEC shifted its focus toward the development of the next generation river analysis system called HEC-RAS, with the first version released in 1996. Sediment was added to HEC-RAS with the version 4.0 release in 2008. A two-dimensional version of HEC-RAS with sediment was released in 2021.

Two-dimensional hydrodynamic numerical models came into production use in the 1960s and two-dimensional sediment transport models followed, with Sediment II (Ariathurai and Krone 1976) among the first. Modifications of Sediment II, including an addition of noncohesive transport, preceded its incorporation into the USACE Tabs-2 numerical modeling system (Thomas and McAnally 1985). Similar progress was being made at the Hydraulics Research Station, Wallingford, UK (now HRS Wallingford). Comparable models were developed in subsequent years at other research institutions worldwide.

Practical three-dimensional numerical models for sediment transport were introduced in the 1980s and 1990s, with Sediment10 (Ariathurai 1982), CH3D-WES (Johnson et al. 1991), Delft3D (Roelvink and Van Banning 1995), and EFDC (Hamrick 1996), followed by a number of others at universities and research centers worldwide. Such numerical models have become a standard tool for predicting navigation channel sedimentation under altered conditions, such as channel enlargement, diversions, and structural alterations. However, simpler mathematical tools continue to be used for smaller projects or as part of initial design stages. The various methods are described in Chapter 5.

ORGANIZATION OF THIS MANUAL

Chapters 2 and 3 provide necessary background information for Chapters 4 to 6. Chapter 2 presents a mainly qualitative overview of channel sedimentation processes—the forces and sediment characteristics that cause sediment transport, deposition, and erosion. That material can be treated as a review for experienced sedimentation engineers. It also describes the significant temporal and spatial scales for channel sedimentation issues, which provide context for the solution method practices described in Chapter 6.

Chapter 3 describes navigation channels, their classifications, and typical sedimentation issues. It can be treated as a review by experienced navigation engineers.

Chapter 4 addresses solutions/remedies to channel sedimentation problems with a recommended classification system for remedies and brief discussions and examples of some remedies. It presents the concepts of sustainability and holism in sediment management, including organized initiatives such as Engineering with Nature and Envision, and how they apply to channel sedimentation.

Chapter 5 categorizes channel sedimentation predictions methods as field observations, desktop analyses, physical (scale) modeling, and numerical modeling and describes each with examples. The final section recommends methods to express and reduce uncertainty in predictions from those methods. It provides a base of information upon which Chapter 6 is built.

Chapter 6 documents our recommendations for best practices in using the prediction methods described in Chapter 5. The chapter emphasizes desktop methods and numerical models used in combination with field observations. Good physical modeling practice parallels the other methods but is not addressed here.

Appendix A is a glossary of terms significant to this manual's content. Appendix B contains case histories that illustrate typical navigation sedimentation problems and the prediction methods used to evaluate remedies. Appendix C provides a decision tree structure to facilitate prediction method selections. Appendix D is a list of acronym definitions.

Standard units in the US navigation engineering field are customary units of the foot-pound-second system; however, sedimentation engineering typically employs the international system (SI) meter–kilogram–second system. Therefore, engineers working in navigation channel sedimentation issues must be accustomed to moving freely between the two systems as the need arises. This manual follows that practice.

TERMINOLOGY

George Bernard Shaw is reported to have said, "England and America are two countries separated by a common language." The same might be said of inland and coastal navigation practitioners and inland and coastal sedimentation engineers. This manual attempts to address these four realms of knowledge using the language of all four. The authors hope that a liberal use of definitions and the glossary in Appendix A will both facilitate reading of the manual and promote an understanding of processes that are common to all environments.

REFERENCES

Ariathurai, R. 1982. *Two and three dimensional models for sediment transport.* Lafayette, CA: Resource Management Associates.

Ariathurai, R., and R. B. Krone. 1976. "Finite element model for cohesive sediment transport." *J. Hydraul. Div.* 102 (3): 323–338.

Burroughs, T. 2020. *Navigation update: American Association of Port Authorities,* AAPA virtual meeting. Washington, DC: US Army Corps of Engineers.

CTH (Committee on Tidal Hydraulics). 1971. *Estuarine navigation projects.* Tech. Bulletin No. 17. Vicksburg, MS: US Army Corps of Engineers.

Fatheree, B. H. 2004. *The first 75 years: History of hydraulics engineering at the Waterways Experiment Station.* Vicksburg, MS: US Army Engineer Research and Development Center.

Hamrick, J. M. 1996. *User's manual for the environmental fluid dynamics computer code.* Special Rep. in Applied Marine Science and Ocean Engineering (SRAMSOE), No. 331. Williamsburg, VA: College of William and Mary, Virginia Institute of Marine Science.

Hudson, R. A., F. A. Herrmann Jr., R. A. Sager, R. W. Whalin, et al. 1979. *Coastal hydraulic models.* Special Rep. 5. Vicksburg, MS: USACE, Coastal Engineering Research Center, Waterways Experiment Station.

Johnson, B. H., R. E. Heath, B. B. Hsieh, K. W. Kim, et al. 1991. *User's guide for a three-dimensional numerical hydrodynamic, salinity, and temperature model of Chesapeake Bay*. Tech. Rep. HL-91-20. Vicksburg, MS: USACE, Waterways Experiment Station.

Kruse, C. J., R. Farzaneh, B. Glover, J. E. Warner, et al. 2021. *A modal comparison of domestic freight transportation effects on the general public: 2001–2019*. Fayetteville, AR: University of Arkansas.

MARAD (Maritime Administration). 2021. *Maritime transportation system (MTS)*. Washington, DC: US Department of Transportation.

Mehta, A. J. 2023. *An introduction to hydraulics of fine sediment transport*. 2nd ed. Singapore: World Scientific.

Mitchell, K. N. 2020. *Quantifying impacts to shipping from vessel draft restrictions*. Washington, DC: American Association of Port Authorities.

Roelvink, J. A., and G. K. F. M. Van Banning. 1995. "Design and development of DELFT3D and application to coastal morphodynamics." *Oceanogr. Lit. Rev.* 11 (42): 925.

Rouse, H., and S. Ince. 1957. *History of hydraulics*. Iowa City, IA: Iowa Institute of Hydraulic Research, State University of Iowa.

Schulz, E. H. 1917. Vol. 9, *The passes of the Mississippi river. Their description, physical characteristics, development, and commerce*, 135–169. Professional Memoirs, Corps of Engineers, United States Army, and Engineer Dept. at Large. Alexandria, VA: Society of American Military Engineers.

Smith, T. P. 2019. *USACE update for the American Association of Port Authorities*. Washington, DC: US Army Corps of Engineers.

Teeter, A. M. 1989. *Effects of Cooper River rediversion flows on shoaling conditions at Charleston Harbor, Charleston, South Carolina*. TR HL-89-3. Vicksburg, MS: USACE, Waterways Experiment Station.

Thomas, W. A., and W. H. McAnally. 1985. *User's manual for the generalized computer program system, open channel flow and sedimentation*. TABS-2, IR HL-85-1. Vicksburg, MS: USACE, Waterways Experiment Station.

USACE (US Army Corps of Engineers). 1995. *HEC-6 reservoir sediment control applications*. TP-148. Davis, CA: USACE, Hydrologic Engineering Center.

USACE. 2017. *Hopper dredge recapitalization analysis*. Washington, DC: USACE.

USACE. 2021. *Dredging Information*. Washington, DC: USACE. https://publibrary.planusace.us/#/series/Dredging%20Information. Accessed December 3, 2021.

USACE. 2022. *Final environmental impact statement for the proposed mid-Barataria sediment diversion project, Plaquemines Parish, Louisiana*. New Orleans: USACE.

Vanoni, V. A. 1975. *Sedimentation engineering*. Reston, VA: ASCE.

Verified Market Research. 2022. "Dredging market size, share, trends, opportunities and forecast." https://www.verifiedmarketresearch.com/product/dredging-market. Accessed March 22, 2022.

CHAPTER 2
CHANNEL SEDIMENTATION PROCESSES

Reliable solutions to, and accurate predictions of, navigation channel sedimentation require having a thorough knowledge of sediment transport processes, including the hydrologic and hydraulic forces driving these processes. This chapter provides an overview of significant waterway sedimentation processes. A more complete description can be found in ASCE MOP 54 (Vanoni 1975) and MOP 110 (Garcia 2008), example references cited in the following, and the scientific and engineering literature, notably the *Journal of Waterway, Port, Coastal and Ocean Engineering* and the *Journal of Hydraulic Engineering*.

DOMINANT/DRIVING FORCES

Broadly speaking, the principal external forces of interest are water, wind, and gravity. In riverine locations, gravity is the dominant external energy force, and the hydrodynamic equations need only to translate that force into flow velocity and boundary shear. The computation is often uncomplicated by variations in fluid densities or wind energy. In coastal locations, wind and tides may be the dominant external energy forces of interest. However, water properties of density and temperature may vary in the vertical and horizontal directions.

Also broadly speaking, sediment transport in rivers, lakes, estuaries, and coastal settings involves the advection and diffusion of sediment originating from freshwater and/or marine sources, and the bed-water exchange processes of erosion and deposition. Transport is a function of the physical forcings acting upon the sediment, including the fluid forces of barotropic flow (water density a function of pressure only), baroclinic circulation (water density a function of pressure, temperature, and constituent concentrations), freshwater flow, waves, and winds. In addition to impacting transport directly, some of these forcings may also have secondary effects via nonlinear interactions with the others. Anthropogenic activities, primarily vessel traffic, can also directly impact sediment transport.

Barotropic flow phenomena typically occur in river and coastal/ocean settings, although they also occur in relatively large inland freshwater systems such as the Great Lakes in the United States and Canada. Barotropic processes include river and lake surface slopes, periodic astronomic tides, and episodic events, such as wind-, wave-, and atmospheric pressure-induced setup/setdown and seiche, all of which produce flows driven by water-level gradients.

Barotropic processes are modified by system geometry, morphology, and earth's rotation. For instance, in coastal areas, the astronomic ocean tide becomes distorted by waterbody friction and geometry, including the relatively shallow water depths, resulting in amplification or damping and asymmetries (in durations and amplitudes) between flood (period of inflowing currents) and ebb (period of outflowing currents). In general, sediment transport follows the direction of the asymmetry in tidal currents, with flood-dominant systems typically tending to import sediments from locations seaward, and ebb-dominant systems tending to export sediments to locations seaward (Friedrichs and Aubrey 1988). In addition, both tidal distortion and the spatially variable and oscillatory nature of tidal currents give rise to several residual sediment transport mechanisms, grouped into what is referred to as "tidal asymmetries and lag effects" (Gatto et al. 2017). Finally, use of a fixed coordinate system on the earth's surface may require a consideration of the Coriolis effect, in which large water body flows appear to veer to the right (in the northern hemisphere) because of earth rotation. This effect becomes more pronounced with increasing latitude.

Note that tidal systems may have little net water mass exchange while simultaneously having significant net sediment mass exchange. This is because the bed tractive force usually varies as the square of the velocity, and therefore, it is possible to have a symmetric net mass balance together with an asymmetric net force balance with resulting net sediment transport. Flows also vary spatially with ebb and flood differences.

Freshwater inflows in waterways may be associated with river flow, groundwater discharge, stormwater outfalls, industrial and municipal discharges and so on. These may vary seasonally and episodically (in response to rainfall runoff during storm events). In addition to delivering sediment, freshwater flow can also impact sediment dynamics through several mechanisms:

- Flood flows extending into overbank areas in a floodplain may deposit sediment in some areas and erode sediment in others (e.g., Gaines et al. 2021).
- High freshwater flow during storm events can increase bed shear stress, increasing watershed yield, increasing sediment transport, and causing bed and bank erosion with downstream deposition (e.g., Ralston et al. 2013).
- Sediment-laden freshets flowing into lakes can create density currents that carry sediment far into the lake (e.g., Lohrenz et al. 2004).

River flows often display a hysteresis effect between rising and falling stages and their nonunique corresponding discharges. Similarly, sediment discharge may exhibit a hysteretic loop when plotted against stage, sometimes counter to the direction of the water discharge hysteresis (e.g., Malutta et al. 2020). River flows may also exhibit helical flow patterns in bends with surface water forced to the outside of the bend and bottom water forced toward the inside of the bend (e.g., Engelund 1974). The helical flow tends to cause deposition on the inside of bends and erosion on the outside with midchannel deposition/erosion patterns varying with discharge (Franco 1971). This, in turn, leads to the classical morphologic evolution of the meandering river, whereby river bends grow progressively more sinuous until the river avulses and isolates the most exaggerated bends, creating oxbow lakes.

Baroclinic circulation is typically seen in coastal, estuarine, and lake/reservoir settings. For instance, in an estuary, it is usually induced by an along-channel density (salinity) gradient resulting from the mixing of freshwater from the head-of-tide and saline water from the ocean. The resulting longitudinal pressure gradient induces residual near-bottom currents directed up-estuary and near-surface currents directed down-estuary within the zone of salinity intrusion. Because of its dependence on the longitudinal salinity gradient, which, in turn, depends on freshwater inflow, the strength of estuarine circulation increases as river flow increases (e.g., Festa and Hansen 1978, Simmons 1969). In combination with a vertical gradient in suspended sediment concentrations (typically increasing from the surface to the bottom of the water column), estuarine circulation induces an up-estuary flux of sediment that extends up to near the limit of salinity intrusion. This results in the formation of estuarine turbidity maxima (ETM), zones of elevated suspended sediment concentrations, and enhanced sediment trapping, often co-located with the salt front (Festa and Hansen 1978, Sanford et al. 2001).

Freshwater flow to coastal areas contributes to baroclinic and barotropic circulation in several ways: High freshwater flow can reduce estuarine salinity intrusion (and thereby impact circulation and the ETM) and vice versa during low freshwater flow periods (e.g., Mathew and Winterwerp

2020). High freshwater flow can impact tidal asymmetries by altering the relative magnitudes of peak ebb and flood currents (e.g., Mathew and Winterwerp 2020). For instance, a system otherwise flood-dominant during low-average river flow conditions may become ebb-dominant at sufficiently high river flows, thus altering the direction and magnitude of net suspended sediment transport.

Waves, either generated offshore or generated locally by winds, may act episodically in association with wind events, storms, hurricanes, and so on. Wave impacts on sediment transport are typically the greatest nearshore, within the wave-breaking or surf zone, and result in either erosion or deposition, depending on wave steepness, turbulence generation, wave-generated currents, and sediment characteristics. Offshore of the surf zone, near-bed orbital velocities and pressure fluctuations generated by surface waves can result in enhanced bed shear and normal stresses and suspend bed sediment. The suspended sediment may be transported by ambient currents and nonlinear wave orbital velocities.

In addition to generating wind-waves, winds can also induce circulation in the upper layers of the water column and thus affect transport and mixing of suspended sediments.

Vessel traffic may have a profound effect on channel sedimentation in narrow waterways and shallows adjacent to navigation channels. Propwash generated by ship propellers can extend to the bed and erode sediment within the navigation channel, with the location and magnitude of scour dependent on factors such as vessel draft, power applied by the propellers, traffic and so on (Craig et al. 2023). For instance, in Newark Bay, located in New York Harbor and home to the Port of New York and New Jersey, Clarke et al. (2015) found variable patterns of resuspension depending on vessel type and activity—tugs pushing barges did not induce resuspension, whereas tugs assisting ships in rotating and docking maneuvers appeared to induce resuspension. Similar variability was also noted for other vessels (deep-draft versus shallower draft car carriers). Spatial patterns of erosion and deposition within navigation channels can also correlate with the locations of vessel traffic and maneuvering. For instance, in Newark Bay, USACE (2006) found that transiting vessels tend to prevent net deposition (via propwash induced erosion) in the center of the channel and in areas of vessel maneuvering (e.g., a bend in the navigation channel), with deposition occurring primarily along the toes of the navigation channel and in relatively quiescent port channels. Vessel bow, stern, and transverse waves may suspend bed sediment and erode banklines, both submerged and subaerial (e.g., Clarke et al. 2015, Maynord 2003).

Icing of waterways and banks can increase erosion through higher flow velocities under ice cover, ice floe gouging, and freeze–thaw cycles that cause soil heaving and piping (e.g., Gatto 1995).

The integrated effect of these various transport processes often produces a spatially and temporally variable distribution of relevant forcings. For example, in Newark Bay, sediment transport is dominated by barotropic processes during typical weather conditions but dominated by locally generated wind-waves during episodic storm events (Shrestha et al. 2014). However, in the Lower Passaic River, which is a tributary of Newark Bay, sediment transport is dominated by barotropic and baroclinic circulation during low-average river flow conditions but dominated by fluvial processes during above-average river flow conditions (Mathew and Winterwerp 2020). Therefore, the assessment and mitigation of problems related to sediment management requires an analysis of the several relevant hydrodynamic forcings and their impacts on sedimentation. Consequently, predictions of sedimentation responses to solutions or changed conditions must thoroughly consider which forcing processes (and interactions of those processes) are significant and how they should be represented.

SEDIMENT SOURCES AND SINKS

Channel sediments may have alluvial and/or marine sources. Alluvial sediments are derived from terrestrial processes such as human activities in the watershed, weathering and abrasion of rock, entrainment of silts and clays from overland flow, and bed and bank erosion in waterways. Sediment sinks in alluvial systems range from long-term depositional bars formed as a result of migratory channels to floodplains to distributaries to downstream lakes and coastal waters. Navigation channels that require maintenance dredging become permanent sediment sinks if the dredged sediment is placed in upland or confined areas.

Marine sediments may originate both from watersheds connected to the ocean and from erosion of the marine environment, including islands, beaches, and offshore substrate (e.g., reefs and bars). Marine sediments are transported both along and across coastlines by the action of waves and currents (driven by breaking waves, tides, winds, and local tributary flows). In marine systems, transport of sediment offshore to the deeper continental shelf constitutes permanent sediment sinks and coastal lagoons can act as long-term sediment sinks.

The source, transport, and eventual sink of sediment represent a complex interplay of long and short-term processes (see the following section) that must be understood to predict future behavior. In general, prediction of sedimentation in navigation channels involves shorter term processes on the order of seasonal to annual timescales that deliver sediment from source areas to the channel. Nevertheless, longer-term processes must be considered in projects designed for 50-year life spans and most often employed for far longer periods. Numerous navigation channels and their control structures in the United States are well over 100 years old and elsewhere they can be much older.

SEDIMENTATION SCALES

Sedimentation processes occur at a variety of temporal and spatial scales. For example, fine sediment (micrometer spatial scale) flocculation happens on the order of minutes, whereas landscape alterations such as erosive mountain leveling and basin infilling take millennia and cover continents. Figure 2-1 illustrates where some sedimentation phenomena and their driving forces fit on that scale of processes. The axes are not linear, but the processes tend to line up along the (nonlinear) dashed line, indicating a correlation between space and time scales except for outliers such as dust storms and individual grain wear from abrasion.

Figure 2-1 reveals disconnects between some processes and navigation channel projects. For example, public works projects are typically designed for a 50-year life and channels are designed with depths specified to the nearest foot (0.3 m). Navigation engineers must deal with channel sedimentation issues of about 1 m in the vertical to kilometers in the horizontal plane and months to decades of time—approximately in the middle of Figure 2-1. In contrast, processes driving those issues occur on

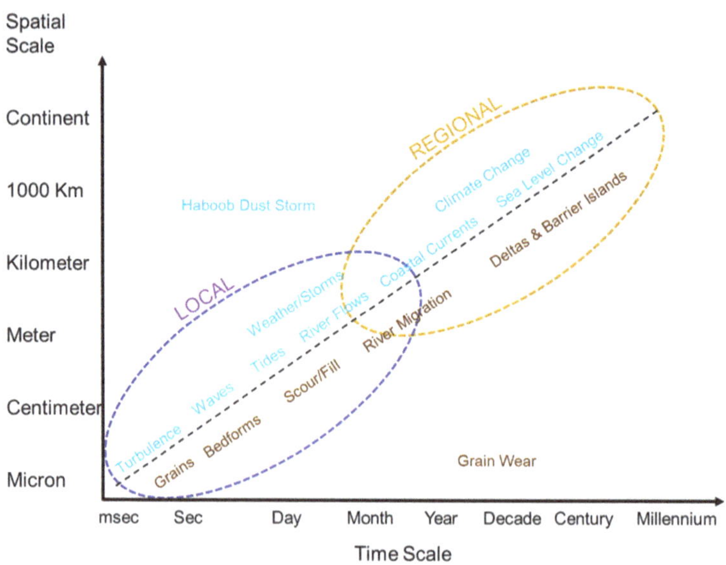

Figure 2-1. Time and space scales of some natural forces (blue, above dashed line), and sedimentation processes (brown, below dashed line).

both smaller scales—micrometers to meters and seconds to days—and larger scales—a few to thousands of kilometers and decades to centuries. These disconnects between sediment processes and navigation channel projects affect the design and rehabilitation of navigation projects. Therefore, they affect the selection of solutions (dikes, dredging, and so on) and the practice of making sedimentation predictions. A prediction that accounts for very small-scale processes such as continuous flocculation can become unwieldy or even erroneous if extrapolated to years or decades. A prediction that accounts for medium-scale processes such as observed deposition rates may miss a drastic change induced by small-scale flocculation or large-scale dune migration. For example, Byrnes et al. (2007) identified shoal migration of about 3 km over a period of 50 years at the mouth of the Columbia River between Washington and Oregon. Such shoals may have no channel impact for decades yet impose a large dredging burden within a few months if they migrate into the designated navigation channel.

Regional and continental scale sedimentation can be considered geomorphological processes. On these scales, landforms erode, rivers may change course as they carry eroded sediment downstream in various planform patterns, lakes and estuaries trap sediments and gradually fill, coastal inlets close and open, littoral transport carries sediments along coasts, and coastlines advance and retreat. These geomorphic processes constitute a navigation project's setting and must be included in the engineering analysis of navigation channel sedimentation. For a thorough discussion of geomorphological processes, see Schumm and Harvey (2007).

For these reasons, operations of navigation projects tend to be dominated by processes acting on the order of days to months; however, future planning requires a consideration of both larger and smaller scales to prevent budget-breaking surprises. Designing and modeling solutions for engineering scales must account for processes at much smaller and much larger scales.

SEDIMENT PROPERTIES

Sediments are composed of both fine (clays and silts) and coarse (sands and gravels) materials. The commonly used metric for defining sediment size is the grain size equivalent diameter expressed in various dimensions or sieve sizes, as shown in Table 2.3 of Garcia (2008) and Table 2.1 of Vanoni (1975). Fines are defined as either particles smaller than sand (<62 μm) or as the sediment passing through a No. 200 sieve (<75 μm) depending on the application. In general, larger particles settle faster and require more flow energy for transport. Flocculated fine sediment is an exception, because floc density decreases with size and settling velocity increases slowly with size, if at all. Other properties such as organic content can have important influences on sediment transport and are addressed subsequently.

Density is used alongside particle size distribution to describe a sediment bed. Bed density is usually expressed as a dry density, which is the mass of the sediment particles divided by the total volume of sediment particles and pore spaces between the particles. Other bed properties of interest include organic content and rheologic measures. Variation of all properties with depth into the bed may be significant. Many standard references exist for defining sediment bed properties. Mehta (2018), Van Rijn (1993), and Yang (1996) provide summaries.

Sediment can be classified as cohesive or noncohesive. Cohesive sediment behavior is affected by interparticulate attractive forces that cause them to aggregate into flocs containing hundreds to millions of individual mineral grains. Cohesive sediments are typically composed of clay-sized (<2 μm) and silt-sized (<62 μm) particles. Often, these fine sediments are mixed with organic matter and, sometimes, quantities of very fine sand; however, the interparticulate forces are still significant. A sediment mixture with a clay fraction greater than about 10% is usually sufficient for the sediment to exhibit cohesive properties (van Rijn 1993). High-concentration cohesive sediment suspensions may exhibit non-Newtonian behavior when in a fluid-like state (Mehta 2018, Winterwerp and Van Kesteren 2004).

Noncohesive sediments are sand and gravel-sized material (>62 μm) plus coarse silts where the particles exert only weight, friction, and impact forces on each other. Overall, noncohesive sediment transport properties are much better understood than cohesive sediment properties because of the difficulty of quantitatively predicting the role of interparticulate forces in cohesive sediment. Van Rijn (1993), Yang (1996), and Garcia (2008) provide excellent descriptions of noncohesive sediment transport.

SEDIMENT DYNAMICS

A schematic of the processes involved in the transport of cohesive and noncohesive sediment in the water column and the sediment bed is shown in Figure 2-2, in which C indicates the suspended sediment concentration profile, U indicates the velocity profile, τ_b is the bed shear stress, and u_* is the shear velocity, as expressed in Equation (2-1). The concentration profile reflects increased sediment concentration with depth caused by particle settling, balanced by an upward diffusive flux of sediment associated with turbulent vertical velocity fluctuations coupled with a vertical sediment

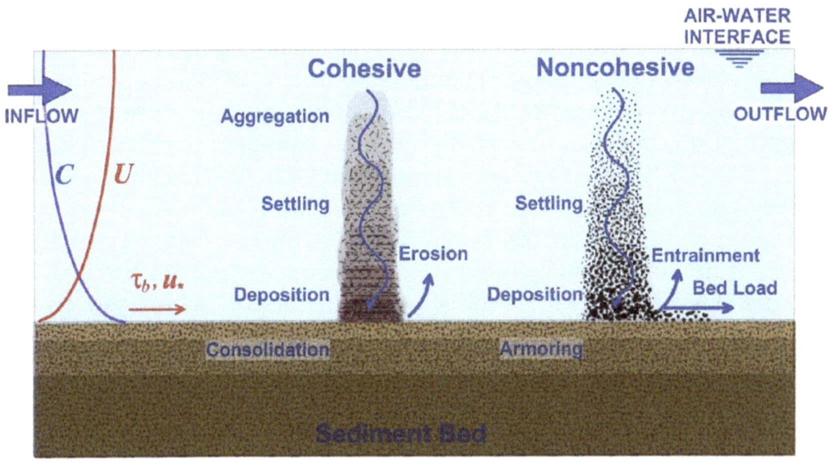

Figure 2-2. Conceptual schematic of sediment transport processes. Source: Shrestha et al. (2014). Used with permission.

concentration gradient. Note that the velocity profile decreases to zero at the sediment bed, consistent with the no-slip condition

$$u_* = \sqrt{\frac{\tau_b}{\rho}} \tag{2-1}$$

where ρ is the water density.

Sediment particle settling velocity for small particles can be expressed by Stokes law, as shown in Equation (2-2), or empirical methods such as given by Cheng (1997). For larger particles, the settling velocity can be calculated using the drag law, which is approximated by

$$W_s = \frac{2(\rho_p - \rho)gr^2}{9\mu} \tag{2-2}$$

where

ρ_p = Density of the sediment particle,
g = Acceleration of gravity,
r = Radius of the particle, and
μ = Fluid dynamic viscosity (Cheng 1997).

Moving water, whether because of river flow, tides, or waves, exerts a shear stress on the sediment bed. The shear stress, if large enough, may mobilize or erode sediment from the sediment bed and suspend it into the water column. Local advection and diffusion in the water column control the transport of suspended sediment particles throughout the water column. Advection moves suspended sediment according to the local water velocity, whereas turbulent diffusion spreads sediment at rates dependent on concentration gradients and turbulence (e.g., Hayter et al. 2014, Yang 1996). Further, sediment [both noncohesive and cohesive (Perkey and Smith 2019)] may also be transported as bed load—moving in proximity to, and in frequent contact with, the bed. Noncohesive sediment moving as suspended load still exchanges with the bed, but turbulent sweeps and bursts transport sediment pulses between the bed and the water column. When sediments move into lower-energy regions where turbulence and stresses are not high enough to keep them moving, the sediment settles and deposits. Yang (1996), Van Rijn (1993), McAnally and Mehta (2000), and others provide detailed descriptions of these processes.

Short-period waves affect sediment transport by exerting both shear and normal stresses on the sediment bed. Alternating direction and magnitude wave shear stresses combine nonlinearly with quasi-steady flow-induced shear stresses (e.g., Pascolo et al. 2018). Normal (pressure) stresses exerted by a wave can alternately fluidize and compact a porous sediment bed (Qi et al. 2019), thereby temporarily enhancing and inhibiting transport. Breaking waves convert tremendous amounts of energy into sediment transport potential and drive longshore and cross-shore transport of all sediment sizes from fines to cobbles.

Transport under the influence of waves establishes a different mode of energy exertion on the bed. In addition to interaction with the ambient current, waves generate turbulence and three-dimensional motion of water particles that is not uniform in space and time. Waves influence currents and sediment in the shallow water environment (depth less than half the wavelength) where a wave begins to "feel" the bottom. Waves change direction through refraction, diffraction, and reflection interactions with the bed, ambient current, and structures, and thus may modify the main flow direction in shallow water environments.

As shallow water waves approach the shoreface, they change shape (becoming skewed and asymmetric), shoal (increase in height), and break (e.g., spilling, plunging, or collapsing) in shallow water, causing significant turbulence and energy dissipation on the bed. As waves approach the nearshore and coastal entrance channels, sediment transport increases over the adjacent beaches and shoals because of the dissipation of breaking waves and subsequent turbulence ejecting suspended sediment into the water column. The mobilized sediment transits the nearshore via longshore currents and can enter tidal inlets, resulting in sedimentation near the entry point or channel reaches of slower ambient (tidal) current velocity within the channel.

Larson et al. (2003) calculated the transport potential of a wave-induced alongshore current with a suspended sediment plume of a commonly observed fine sand-sized particle and concluded that sediment mobility decreases as it exits the shallow nearshore and enters a deep tidal (navigation) channel. Sediment trapping within the entrance channel is directly related to the sediment pathway connecting the adjacent nearshore environment to the channel as driven by tidal- and wave-induced longshore currents (Beck et al. 2020). In barrier islands, this feature is referred to by geomorphologists as a flood (marginal) channel, which is often associated with erosion of adjacent shorelines and a direct contributor to entrance channel sedimentation (Beck and Wang 2019).

Over longer timescales of months to years, unconstrained nearshore environments and increased mobility of sediment at coastal entrance channels can change morphology and velocity characteristics of navigation channels. Extratropical and tropical storms can accelerate this natural process to occur in the span of days to weeks. A pronounced morphodynamic change of coastal entrance channels is typically greater at unstructured, natural coastal entrance channels, as opposed to those with jetties and other navigation structures (FitzGerald et al. 2000).

COHESIVE SEDIMENT DYNAMICS

Cohesive sediment erosion, transport, aggregation, settling, deposition, and consolidation depend on the inherent physicochemical characteristics of the sediment and their interaction with the water and its flow field.

Aggregation (flocculation) is the formation of flocs composed of primary sediment particles. The processes influencing aggregation are destabilization and collision (Lick and Huang 1993, McAnally and Mehta 2002). When the layer of ions surrounding each suspended sediment particle is compressed by divalent ions, the sediment particles are destabilized and van der Waals forces of attraction predominate, creating a condition that is conducive to aggregation if particles collide. Collision occurs as a result of Brownian motion, internal flow shear, and differential settling. As particles aggregate and form larger particles, the settling speed of those particles typically increases.

The settling velocity of cohesive sediments depends on the individual or aggregate effective particles' size, shape, and density (McAnally and Mehta 2002). Further, at high sediment concentrations, hindered settling occurs (i.e., the settling velocity decreases). Fluid mud is a high concentration suspension, usually separated from the overlying, lower concentration layer by a lutocline, a strong vertical concentration gradient, when sediment concentrations exceed 10 kg·m^{-3} (McAnally et al. 2007).

As aggregates settle toward the sediment bed, near-bed turbulence controls whether the particles deposit to the bed or are broken up and/or re-entrained into the water column. Near-bed turbulence is characterized by a "probability of deposition," which is defined as the probability that particles reaching the bed actually stick to the bed. The deposition rate is thus a function of the aggregate settling velocity, near-bed concentration, aggregate strength, and flow stresses, each of which varies probabilistically (Krone 1962, Letter and Mehta 2011, Mehta et al. 2014).

Deposited aggregates consolidate under the weight of additional sediment depositing on top of them. Thus, the void ratio of the bed decreases with the depth of the sediment. This results in an increase in the density and shear strength with depth into the bed (Winterwerp and van Kesteren 2004, Van Ledden et al. 2004).

Sediment suspension (erosion) from the bed depends on the bed shear stress induced by the flow and the resistance of the sediment bed to erosion (i.e., the critical shear stress for erosion of the bed sediments). In coastal channels, bed shear stresses can be caused by the combined effect of waves and currents (van Rijn 1993, Winterwerp et al. 2021).

Sediment erosion may occur as floc erosion, surface erosion, mass erosion, and liquefaction (Mehta 2018). Floc and surface erosion occur at low-to-moderate values of the excess shear stress where currents are low to moderate. Mass erosion refers to the erosion of larger clumps or layers of material when the bed shear stresses are much larger than the critical shear strength of the bed. Liquefaction occurs when the bed pore pressure, increased by surface waves, exceeds the shear strength of the bed.

Cohesive sediment transport is usually calculated by using the advection–diffusion mass conservation equation with bed erosion and deposition computed as source and sink terms, respectively. Typical forms of source-sink equations are found in MOP 110 (Garcia 2008), and others can be found in the recent literature and numerical models mentioned in Chapter 5, "Prediction Methods."

NONCOHESIVE SEDIMENT DYNAMICS

Noncohesive sediments are transported as bed load or suspended load depending on flow-induced stresses and the settling velocity of the sediment particles. The threshold at which motion of bed sediment begins (incipient motion) owing to a combination of lift and drag forces is characterized by a critical shear stress, τ_c, which depends on grain size, weight, and angularity plus bed slope.

As the flow's shear stresses exceed the critical shear stress of a noncohesive sediment bed, the individual particles begin to move. When the shear stress exceeds the critical shear stress by only a small amount, particles move by rolling and bouncing along the bed—bed load transport. Bed load transport occurs when the bed shear velocity exceeds the critical shear velocity of the bed sediments, $u_c = \sqrt{\tau_c/\rho}$, but is lower than the settling velocity of the sediment particles (van Rijn 1993).

When the bed shear velocity exceeds the settling velocity of the sediment particles, the sediment particles can be lifted from the bed to a level where upward turbulent forces are higher than the submerged weight of the particles and the particles are suspended into the water column. Suspended load includes bed material load plus wash load. Wash load is distributed more uniformly in the water column and transported by the mean velocity but does not appear in the bed. Nevertheless, wash load cannot be ignored, because wash load in one reach may deposit to the bed and become bed material load in another reach or at a lower discharge.

Although suspended load usually transports at about the same velocity as the surrounding fluid, bed load transports at lower-average velocities. As with cohesive sediment, when a lower shear environment is encountered that is lower than the critical shear stress of the particles, the particles will settle to the bed.

Noncohesive sediment beds take shapes as flat (plane) or bedform surfaces—ripples, waves, dunes, or antidunes as a consequence of interaction between flow and sediment characteristics. The bedforms migrate as sediment deposits on one side and erodes from the other.

The four primary noncohesive sedimentation processes are

- Erosion,
- Entrainment,
- Transportation, and
- Deposition.

The processes can be readily identified in the one-dimensional sediment continuity equation

$$\frac{\partial Q_s}{\partial x} + (1 - P)\frac{\partial A_d}{\partial t} = 0 \tag{2-3}$$

where

Q_s = Volumetric sediment discharge,
x = Distance along the channel,
P = Porosity of the bed deposit,
A_d = Volume of sediment deposited on the bed per unit length of channel, and
t = Time.

A fifth process, compaction of the deposits, is secondary to these.

The first term in Equation (2-3) describes the transportation process. The second term quantifies the entrainment and deposition processes. When the first term is zero, the sediment load being transported by the water does not change with distance along the channel. Therefore, the second term must also be zero, and consequently, the only sedimentation process is transportation.

Numerical models use a variety of equations, including the advection–diffusion equation and bed load equations, to calculate noncohesive sediment transport rates. Classical transport equations describe equilibrium transport rates—the noncohesive sediment-carrying capacity of the flow. They include empirical relationships between transport rates and both sediment character (e.g., grain size) and flow intensity (e.g., velocity, shear stress, energy). Hans Einstein (1950) broke new ground with a rigorous mathematical treatment of the physical processes in his equilibrium transport equation. More recently, transport equations have been developed, which represent time-varying transport rates based on live-bed exchange. Many classical equations and more recent equations are found in MOP 110 (Garcia 2008), and others can be found in the professional literature and numerical models mentioned in Chapter 5. The user should understand which is being used and how applicable it is to the situation being evaluated.

OTHER FACTORS

Winnowing of more easily eroded grains from the bed surface may cause armoring to occur—a surface layer of more erosion resistant material. Water quality and biotic activity may affect sedimentation processes. Salinity increases flocculation rates for cohesive sediment up to about two practical salinity units (psu) and may have small effects on aggregate density and strength at higher salinities. Bacteria appear to have mediating effects on fine sediment cohesion, and microbial mats on bed surfaces are known to inhibit erosion in some nutrient rich locations (Edzwald et al. 1974). It has been suggested that bacterial extracellular polymer substance secretions can affect the stability of fluid mud suspensions (PIANC 2009, Pang et al. 2018), slowing dewatering and permitting vessels to safely sail through the suspension, which is the practice of Active Nautical Depth, described in Chapter 4, "Solutions to Sedimentation Problems."

In addition to the self-weight consolidation of cohesive and mixed sediments described previously, several other processes affect the density and erosion resistance of bed sediment. For example, bacterial growth binds sediment grains together and burrowing worms and their wastes decrease bed density. Burrowing can also lead to mixing of sediments over the depth of biological activity and decrease erosion resistance, that is, increase erodibility.

REFERENCES

Beck, T. M., and P. Wang. 2019. "Morphodynamics of barrier-inlet systems in the context of regional sediment management, with case studies from west-central Florida, USA." *Ocean Coastal Manage.* 177 (July): 31–51.

Beck, T. M., P. Wang, H. Li, and W. Wu. 2020. "Sediment bypassing pathways between tidal inlets and adjacent beaches." *J. Coastal Res.* 36 (5): 897–914.

Byrnes, M. R., S. F. Griffee, and H. R. Moritz. 2007. "Engineering activities influencing historical sediment transport pathways at the Columbia River Mouth, WA/OR." In *Coastal Sediments 07*, edited by N. C. Kraus and J. D. Rosati. Reston, VA: ASCE.

Cheng, N. A. 1997. "Simplified settling velocity formula for sediment particles." *J. Hydraul. Eng.* 123 (2): 149–152.

Clarke, D., K. J. Reine, C. Dickerson, C. Alcoba, et al. 2015. *Sediment resuspension by ship traffic in Newark Bay, New Jersey.* ERDC/EL TR-15-1. Vicksburg, MS: US Army Engineer Research and Development Center.

Craig, P. M., J. Y. Jung, Z. Mausolff, L. A. Bastidas, et al. 2023. "Modeling sediment resuspension and transport processes induced by propeller wash from ship traffic." *J. Hydraul. Eng.* 149 (5). https://doi.org/10.1061/JHEND8.HYENG-13229.

Edzwald, J. K., J. B. Upchurch, and C. R. O'Melia. 1974. "Coagulation in estuaries." *Environ. Sci. Technol.* 8 (1): 58–63.

Einstein, H. A. 1950. *The bed-load function for sediment transportation in open channel flows.* USDA Soil Conservation Service, Tech. Bulletin No. 1026. Washington, DC: United States Dept. of Agriculture Soil Conservation Service.

Engelund, F. 1974. "Flow and bed topography in channel bends." *J. Hydraul. Div.* 100 (11): 1631–1648.

Festa, J. F., and D. V. Hansen. 1978. "Turbidity maxima in partially mixed estuaries: A two-dimensional numerical model." *Estuarine Coastal Mar. Sci.* 7 (4): 347–359.

FitzGerald, D. M., N. C. Kraus, and E. B. Hands. 2000. *Natural mechanism of sediment bypassing at tidal inlets.* ERDC/CHL CHETN-IV-30. Vicksburg, MS: US Army Engineer Research and Development Center.

Franco, J. J. 1971. *Development of alluvial streams for navigation.* MP H-71-3. Vicksburg, MS: USACE, Waterways Experiment Station.

Friedrichs, C. T., and D. G. Aubrey. 1988. "Non-linear tidal distortion in shallow well-mixed estuaries: A synthesis." *Estuarine Coastal Shelf Sci.* 27 (5): 521–545.

Gaines, R. A., S. C. Sanborn, W. H. McAnally, and C. M. Wallen. 2021. *Mississippi river adaptive hydraulics model development and evaluation, commerce to New Madrid, Missouri, Reach.* Mississippi River

Geomorphology and Potamology Program, MRG&P Rep. No. 35. Vicksburg, MS: USACE Mississippi Valley Division.

Garcia, M. 2008. *Sedimentation engineering: Processes, measurements, modeling, and practice*. Manuals and Reports on Engineering Practice No. 110. Reston, VA: ASCE.

Gatto, L. W. 1995. *Soil freeze–Thaw effects on bank erodibility and stability*. Special Rep. 95-24. Hanover, NH: USACE Cold Regions Research & Engineering Laboratory.

Gatto, V. M., B. C. Van Prooijen, and Z. B. Wang. 2017. "Net sediment transport in tidal basins: Quantifying the tidal barotropic mechanisms in a unified framework." *Ocean Dyn*. 67 (September): 1385–1406.

Hayter, E., K. Gustavson, S. Ells, J. Gailani, et al. 2014. *Technical guidelines on performing a sediment erosion and deposition assessment (SEDA) at superfund sites*. Vicksburg, MS: Engineer Research and Development Center.

Krone, R. B. 1962. *Flume studies of the transport of sediment in estuarial shoaling processes*. Final Rep. Berkeley, CA: Hydraulic Engineering Laboratory and Sanitary Engineering Research Laboratory, University of California.

Larson, M., N. C. Kraus, and H. Hanson. 2003. "Simulation of regional longshore sediment transport and coastal evolution–the "Cascade" model." In *Coastal Engineering 2002: Solving Coastal Conundrums*, edited by J. M. Smith, 2612–2624. Singapore: World Scientific.

Letter, J. V., Jr., and A. J. Mehta 2011. "A heuristic examination of cohesive sediment bed exchange in turbulent flows." *Coastal Eng*. 58 (8): 779–789.

Lick, W., and H. Huang. 1993. "Flocculation and the physical properties of flocs." In Vol. 42, *Nearshore and estuarine cohesive sediment transport*, edited by A. J. Mehta, 21–39. Washington, DC: American Geophysical Union.

Lohrenz, S. E., G. L. Fahnenstiel, D. F. Millie, O. M. E. Schofield, et al. 2004. "Spring phytoplankton photosynthesis, growth, and primary production and relationships to a recurrent coastal sediment plume and river inputs in southeastern Lake Michigan." *J. Geophys. Res*. 109 (10): C10S14.

Malutta, S., M. Kobiyama, P. L. B. Chaffe, and N. B Bonumá. 2020. "Hysteresis analysis to quantify and qualify the sediment dynamics: State of the art." *Water Sci. Technol*. 81 (12): 2471–2487.

Mathew, R., and J. C. Winterwerp. 2020. "Sediment dynamics and transport regimes in a narrow microtidal estuary." *Ocean Dyn*. 70 (8): 435–462.

Maynord, S. T. 2003. *Ship effects before and after deepening of Sabine-Neches Waterway, Port Arthur, Texas*. TR ERDC/CHL-03-15. Vicksburg, MS: US Army Engineer Research and Development Center, Coastal and Hydraulics Laboratory.

McAnally W. H., C. Friedrichs, D. Hamilton, E. Hayter, et al. 2007. "Management of fluid mud in estuaries, bays, lakes. Part 1: Present state of understanding on character and behavior." *J. Hydraul. Eng.* 133 (1): 9–22.

McAnally, W. H., and A. J. Mehta. 2000. "Aggregation rate of fine sediment." *J. Hydraul. Eng*, 126 (12): 883–892.

McAnally, W. H., and A. J. Mehta. 2002. "Significance of aggregation of fine sediment particles in their deposition." *Estuarine Coastal Shelf Sci.* 54 (4): 643–653.

Mehta, A. J. 2018. *An introduction to hydraulics of fine sediment transport*. 2nd ed. Singapore: World Scientific.

Mehta, A. J., A. J. Manning, and Y. P. Khare. 2014. "A note on the Krone deposition equation and significance of floc aggregation." *Mar. Geol.* 354: 34–39. DOI:10.1016/j.margeo.2014.04.002.

Pang, Q. X., P. P. Han, R. B. Zhang, and C. P. Wen. 2018. "Delaying effect of extracellular polymer substances (EPS) on fluid mud consolidation and application for nautical depth." *J. Waterway, Port, Coastal, Ocean Eng.* 144 (3). https://doi.org/10.1061/(ASCE)WW.1943-5460.0000441.

Pascolo, S., M. Petti, and S. Bosa. 2018. "On the wave bottom shear stress in shallow depths: The role of wave period and bed roughness." *Water* 10 (10): 1348.

Perkey, D. W., and S. J. Smith. 2019. *Impact of mud aggregate processes in sediment transport studies*. ERDC/CHL CHETN-VIII-12. Vicksburg, MS: US Army Engineer Research and Development Center.

PIANC (World Association for Waterborne Transport Infrastructure). 2009. Minimising siltation in harbours. PIANC Rep. 102. Brussels, Belgium: PIANC.

Qi, W.-G., C.-F. Li, D.-S. Jeng, F.-P. Gao, et al. 2019. "Combined wave-current induced excess pore-pressure in a sandy seabed: Flume observations and Comparisons with theoretical models." *Coastal Eng.* 147 (1): 89–98.

Ralston D. K., J. C. Warner, W. R. Geyer, and G. R. Wall. 2013. "Sediment transport due to extreme events: The Hudson River estuary after tropical storms Irene and Lee." *Geophys. Res. Lett.* 40 (20): 5451–5455.

Sanford, T. B., S. E. Suttles, and J. P. Halka. 2001. "Reconsidering the physics of the Chesapeake Bay estuarine turbidity maximum." *Estuaries* 24 (2001): 655–669.

Schumm, S. A., and M. D. Harvey. 2007. "Engineering geomorphology." *Chap. 18 in Sedimentation engineering: Processes, measurements, modeling, and practice*, edited by M. Garcia. Reston, VA: ASCE.

Shrestha, P. L., S. H. Su, S. C. James, P. J. Shaller, et al. 2014. "Conceptual site model for Newark Bay—Hydrodynamics and sediment transport." *J. Mar. Sci. Eng.* 2 (1): 123–139.

Simmons, H. B. 1969. "Field experience in estuaries." Chap. 16 in *Estuary and coastline hydrodynamics*, edited by A. T. Ippen. New York: McGraw-Hill.

USACE (US Army Corps of Engineers). 2006. *Geomorphological/geophysical characterization of the nature and dynamics of sedimentation and sediment transport in Newark Bay focusing on the effects related to continued and future federal navigation channel deepening and maintenance*. New York: USACE.

Van Ledden, M., W. G. M. Van Kesteren, and J. C. Winterwerp. 2004. "A conceptual framework for the erosion behaviour of sand-mud mixtures." *Cont. Shelf Res*. 24 (1): 1–11.

Vanoni, V. A. 1975. *Sedimentation engineering*. Manuals and Reports on Engineering Practice No. 54. Reston, VA: ASCE.

van Rijn L. C. 1993. *Principles of sediment transport in rivers, estuaries, and coastal seas*. Amsterdam, Netherlands: Aqua Publications.

Winterwerp, J. C., and W. G. M. van Kesteren. 2004. *Introduction to the physics of cohesive sediment in the marine environment*. Amsterdam, Netherlands: Elsevier.

Winterwerp, J. C., T. Van Kessel, B. S. Van Maren, and B. C. Van Prooijen. 2021. *Fine sediment in open water – From fundamentals to modeling*. Singapore: World Scientific.

Yang, C. T. 1996. *Sediment transport: Theory and practice*. McGraw-Hill series in water resources and environmental engineering. New York: McGraw-Hill.

CHAPTER 3

NAVIGATION CHANNELS AND SEDIMENTATION

CHANNEL CLASSIFICATIONS

Shallow-Draft and Deep-Draft Channels

Navigation channel design width, depth, and alignment depend on vessel size, traffic flow, and factors such as tides, water levels, winds, waves, and currents. Channel depth is determined by the draft and trim of the design vessel, with additional allowances for factors including tides, wave-induced motions, squat, and underkeel clearance (Briggs et al. 2001).

Navigation channels can be categorized by depth: shallow-draft and deep-draft channels. The distinction between shallow-draft channels and deep-draft channels is ambiguous, with 14 or 15 ft of navigable depth as the most commonly used dividing lines. In the United States, shallow-draft channels typically provide 9 to 12 ft of depth and are found in the inland waterway network, the Gulf and Atlantic Intracoastal Waterways, and small ports nationwide. Portions of the Gulf Intracoastal Waterway have an authorized depth of 16 ft but are still considered shallow-draft channels.

Deep-draft channels offer navigable depths of 18 to 50 ft and more. Within the United States, deep-draft navigation projects involve practically all commercial coastal ports, the lower portions of the Mississippi and Columbia Rivers, and a majority of harbors in the Great Lakes and St. Lawrence River system. Although most deep-draft ports are coastal or on the Great Lakes, some are upstream on navigable rivers. For example, the deep-draft ports of Sacramento and Stockton, California, are about 100 mi inland and multiple deep-draft ports are located up to 253 mi upstream on the Mississippi River from the Gulf of Mexico.

Inland and Coastal

Navigation channels may also be classified as inland or coastal channels. The USACE classifies the Gulf and Atlantic Intracoastal Waterways as inland navigation channels, but they are subject to many of the sedimentation processes of a coastal channel.

Inland channels are found in canals, lakes, and rivers and experience variable sedimentation processes because of high and low flow events (and the frequency and duration of those events) from rain and snow precipitation, upstream and downstream flow management, and land use/ land cover changes. Changes in current patterns (speed, direction or duration) and surface waves because of storms or passing vessels can contribute to channel sedimentation issues, including bank and bed erosion.

Most inland navigation channels in the United States are shallow-draft waterways with authorized depths of about 9 ft and widths on the order of 100 to 300 ft. Channel depths may be greater in lakes, large rivers such as the Mississippi River, and the lower end of navigation pools. Channels in rivers typically follow the thalweg so as to minimize dredging. Training structures such as dikes are commonly used to constrain the channel in wide rivers and to scour a channel through shallows. Channels may also have revetments to stabilize the outer bank of a channel bend and fix the bend in place. Lock and dam structures regulate water depths on many inland waterways and facilitate navigation. For example, 10 lock and dams on the Tennessee–Tombigbee (Tenn–Tom) Waterway and 26 on the upper Mississippi provide a series of stair-step pools to enable navigable channels throughout the navigation seasons. Figure 3-1 illustrates the profile of Tenn–Tom Waterway dams and pools.

Rivers sometimes display meandering channels with sharp bends, point bars, and thalwegs that switch from left to right bank and back again or braided channels in which no single channel provides sufficient natural depth for navigation. Each provides challenges to finding sedimentation solutions that fit the site and its processes.

Coastal and estuarine channels may be either shallow-draft or deep-draft. They experience flow variability like those of inland waterways and are also subject to tidal water elevation fluctuations, mean and episodic sea-level changes, impacts from large tropical and extratropical storm surges, and substantial wave-driven currents. Wind wave effects develop both longshore and cross-shore sedimentation transport paths that affect coastal channels. Estuarine channels experience many of the same processes as other coastal channels plus the frequent occurrence of density currents driven by freshwater–seawater interaction.

Coastal navigation channels can be hundreds of feet deep or be barely navigable because of constantly evolving changes in channel or inlet bathymetry and the encroachment of offshore and bay ebb and flood

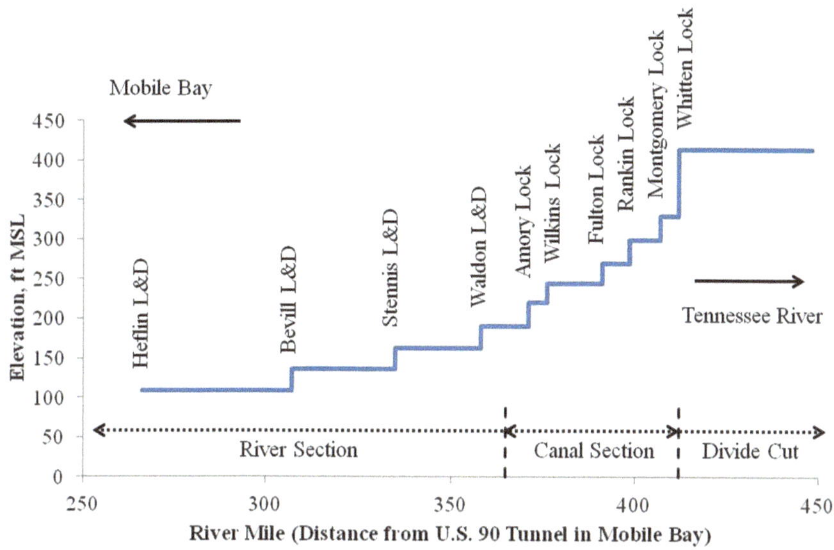

Figure 3-1. Profile of lock and dam structures on the Tennessee–Tombigbee Waterway.

shoals, respectively. There are navigation channels in coastal locations that do not require modification or maintenance as they are naturally deep. However, most coastal navigation channels require training structures and frequent maintenance dredging to maintain navigable depths as a result of channel infilling from either tidal, longshore, or cross-shore sediment transport, or a combination of processes.

Coastal inlet navigation channels are often fixed at specific locations by parallel jetties, channel-perpendicular spur jetties, offshore breakwaters, and groin fields. These features are constructed to minimize the impact of longshore and cross-shore transport on the navigation channel depths. Coastal inlet channels sometimes benefit from natural formations such as headlands, offshore canyons, and diverging nodal points of sediment transport, all of which minimize channel sedimentation impacts.

Channel Configuration and Alignment

Channel configuration and alignment—its geometric features—affect sedimentation processes in multiple, interacting ways. Some major aspects of those configuration are described subsequently.

Navigation channels are typically one of three primary characteristic cross-section types—canal, trench, or fairway—as shown in Figure 3-2. Two types may occur in the same cross section, such as trench type on one side and canal on the other.

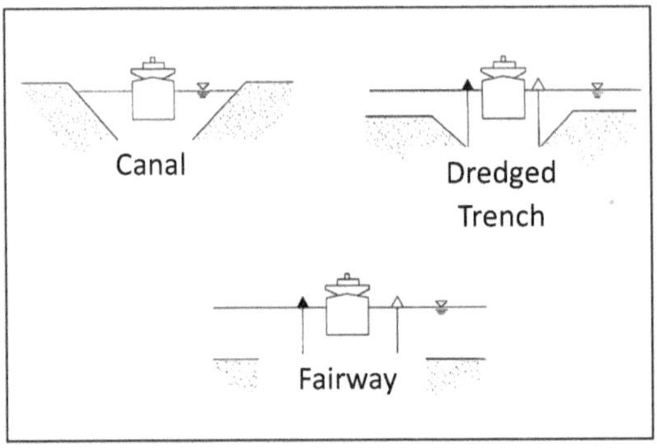

Figure 3-2. Channel types.
Source: USACE (2015).

Channel Reaches

Channel reaches are specific lengths of channels with consistent channel width, depth, and alignment, which are designated so as to distinguish variations in channel dimensions or conditions, as, for example, a channel goes from very deep water through an inlet to shallower water, passes and connects to berthing areas, or bends in alignment because of the natural alignment of a waterway. Channel reaches may also be specified to distinguish channel areas of substantial sediment deposition. This designation of channel reaches may be useful for estimating sedimentation patterns, which, in turn, is used for planning and budgeting dredging or remedial works. Typical generic reach designations from the ocean landward include

- Ocean,
- Bar,
- Jetty,
- Entrance,
- Harbor or port,
- Interior and intracoastal waterway,
- Bay,
- Lock approach, and
- Canal.

Bar and jetty channels are often subject to significant short-period waves propagating in from the ocean and from coastline sediment transport, which can subject them to sedimentation processes unique to their location. Coastline sedimentation processes, like contributing watershed processes, are beyond the scope of this manual, but their effect on bar and jetty channels must be recognized and accounted for. A comprehensive coverage of coastline processes can be found in the *Coastal Engineering Manual* (USACE 2002).

NAVIGATION CHANNEL SEDIMENTATION

Navigation channel sedimentation problems are most often deposition or excessive erosion. Sediment-induced turbidity and sediment bed composition can have detrimental impacts on some habitats. In addition, sediment-induced turbidity may become an aesthetic issue if a community objects to it.

Sediment deposition is a common problem worldwide. Natural water depths are often too shallow to allow safe navigation and dredging them to provide sufficient depth creates an imbalance in the relationship between sediment supply and sediment transport capacity, resulting in deposition. For example, the Mississippi River's Southwest Pass's depths limited vessel draft to about 12 ft before navigation improvements began. Deepened to 48 ft in modern times, the Southwest Pass channel accumulates 10 to 20 million yd^3 of sediment per year that must be dredged to allow safe passage of deep-draft vessels. The issue is described in Appendix B.

Excessive sediment erosion becomes a problem when channel banks, nearby land, structures, or buried utilities are threatened. For example, erosion in Grays Harbor, Washington, created a 60 ft deep hole adjacent to the South Jetty in the late 1960s and threatened to undermine the stone jetty (CTH 1967). Utilities such as pipelines carrying oil and natural gas across channels can potentially be exposed and damaged during high flow events. Channel degradation, localized scour, channel avulsion, and bank erosion can expose buried pipelines resulting in increased hydrodynamic loads that may cause pipeline failure. Such a failure can result in adverse impacts to the environment and ecosystem.

The alignment and configuration (e.g., depth, slope, width) of a navigation channel affects sediment transport in several ways. In general, a deep trench channel cut into a natural alluvial plain or coastal morphology will disrupt the movement of sediment because of reduced current speed and/or wave effects in the deeper water of the channel. The sometimes abrupt reduction in flow speed and shear stresses allows sediment to settle in the channel where waves and currents cannot resuspend it.

A channel aligned with the direction of dominant sediment transport may experience less deposition than one perpendicular to the transport direction. Thus, a channel following the thalweg of a natural waterway will tend to experience less deposition than a channel that crosses the main transport path.

Channels in confined waterways, such as rivers, estuaries, bays, and inlets, often follow an indirect path to take advantage of naturally deeper water or to avoid adverse sailing conditions. Figures 3-3 and 3-4 illustrate both. The Grays Harbor, Washington, bar channel (Figure 3-3) is oriented to point into/away from the dominant wave swell direction for

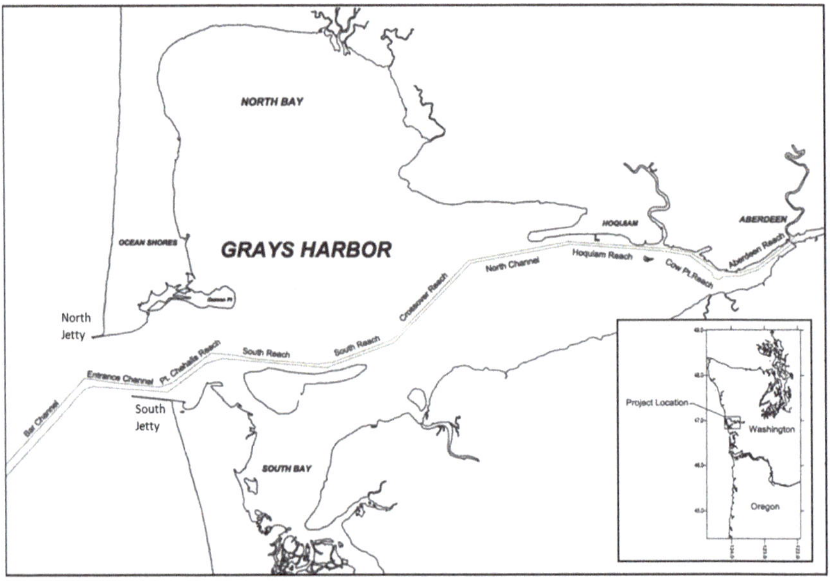

Figure 3-3. Grays Harbor navigation channel.
Source: Hayter et al. (2012).

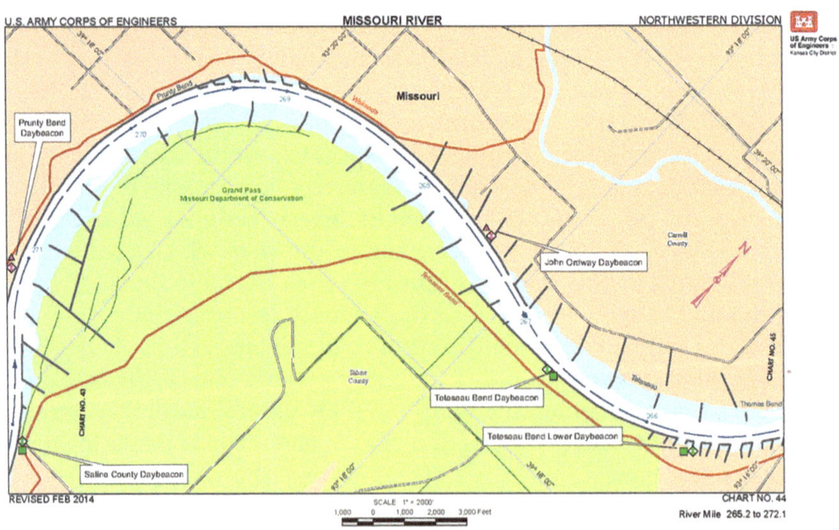

Figure 3-4. Missouri River Navigation Channel from Rulo, Nebraska, to St. Louis, Missouri. Long dashed line shows thalweg.
Source: USACE digital library.

navigation safety. North and South Jetties protect the Entrance Channel from adverse waves and cross currents. The Entrance Channel parallels the South Jetty, where naturally deep water occurs to the extent that erosion has sometimes threatened to undermine the jetty. The channel then veers northward, following deeper water until the Crossover Reach to approach ports and terminals on the north side of the estuary via locality-named reaches.

Figure 3-4 illustrates a sinuous channel in the Lower Missouri River above St. Louis. The channel follows the river thalweg, which is held in place by lateral training dikes (black lines) and revetments as it meanders from side to side of the similarly meandering river. Other structures, such as lateral dikes, chevron dikes, and bendway weirs, perform similar functions (Pokrefke 2012).

Channels approaching and passing through riverine navigation locks, illustrated in Figure 3-5, often present sedimentation challenges, because the pooled water on the upstream side (upper pool) provides a quiet water environment suitable for deposition. The downstream side (lower pool or tailwater) of the dam may be subject to erosion and must usually be protected by concrete aprons and riprap. Locks in coastal waters, such as those on the Gulf Intracoastal Waterway, may lack dams and often experience deposition on both sides of the structure.

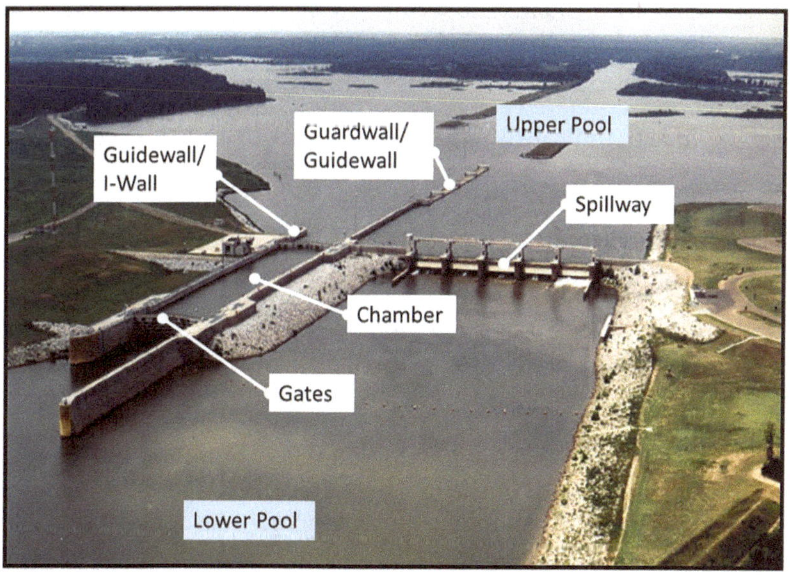

Figure 3-5. Lock and Dam on the Tenn–Tom Waterway.
Source: Adapted from USACE photo.

Harbors, ports, docks, terminals, and marinas plus their connecting channels, anchorages, and turning basins experience the same sedimentation issues as navigation channels plus some unique problems related to their configuration—often in dead-end enclosures—and structure complications. For example, structural pilings beneath docks and quays and hard-edged obstructions create turbulent eddies that increase cohesive sediment flocculation/deposition and decrease flow energy that can cause deposition.

OFF-CHANNEL SEDIMENTATION

Off-channel sedimentation issues include erosion of and deposition to banklines, shallow bars, marshes, and in adjacent waterways influenced by navigation channels. Natural processes usually dominate such waters, and they assume significance to navigation projects only when those projects cause increased deposition or erosion. That cause-and-effect connection can arise in several ways. For example, if a main channel navigation project changes water levels, either steepening or flattening tributary waterway slopes, deposition and erosion patterns in those tributaries will change. Navigation traffic itself may induce sediment resuspension and pumping of sediment laden water into connecting waterbodies. For example, Maynord et al. (2004) investigated side channel deposition related to vessel traffic in the Illinois Waterway System. Coastal inlet channels interrupt longshore transport and may cause updrift and bar accretion and downdrift erosion. These off-channel navigation project-related effects are potentially significant to navigation project management, but their solutions are beyond the scope of this manual.

BANKLINE AND SHORELINE EROSION

Bankline and shoreline erosion can be a source of sediment supply to a channel, potentially increasing channel sedimentation. Inland channel waterway bank erosion contributes to local sediment supply and may be transported far downstream. Eroded estuarine and ocean coastline sediment moves across and along the shoreline via littoral transport and often constitutes a source of channel deposition in harbor inlet channels.

Processes influencing shoreline erosion include storm-event flows with water depths that impinge on the banks of the channel, wave energy impact on channel banks and bank slopes, and landside flows on the channel slope. The effect is to erode the channel banks, adding sediment supply to the channel. Sedimentation occurs on the upstream side of structures such as groins and breakwaters perpendicular/oblique to the channel banks because of the structure blocking the transport of sediments. Erosion usually occurs on the downdrift side. Often, high flows in the channel result in channel side-slope toe erosion and sloughing of material from the banks. Monitoring channel cross-section data using aerial photos and light detection and ranging (LiDAR) digital elevation data in conjunction with soil density data for the channel banks allows computation of the volume (and mass) per unit time of sediment input to the channel. Such estimates are often included as sediment inputs to numerical models to assess their transport and potential deposition downstream.

SEDIMENT QUALITY

Sediment quality can be of concern in two respects—characteristics and contaminants. Fine sediments—silts and clays—are sometimes incompatible with sand and gravel habitat requirements, such as those for salmon, oysters, and mussels. Sea turtles, on the contrary, adapt readily to muddy bottoms. Organic and nonorganic contaminants such as PCBs and metals may adhere to fine sediment and require specialized treatment of the contaminated materials. Dredging equipment and location may be constrained and disposal of contaminated sediments limited to specified places and conditions (EPA 1992). For instance, contaminated sediments may require environmental dredging techniques such as clamshell dredge-buckets that are relatively inefficient (in terms of dredging rates) to minimize dredge/contaminant spillage and resuspension. Contaminated sediments also typically have relatively high disposal costs, requiring treatment and/or disposal in landfills permitted to handle toxic wastes. It is more effective to prevent contamination of the material to be dredged by eliminating the source of the contamination prior to its contact with channel sediment.

In contrast to contaminated sediments, uncontaminated sediments may be dredged using more productive methods such as hydraulic dredging and the dredged material used beneficially for wetland restoration, beach nourishment, and so on (e.g., PIANC 2023).

Episodic high flow events in watersheds often introduce large quantities of noncohesive sediments that are relatively clean, and the resulting effect is to reduce the contaminant concentration in the water column and the bed of the channel owing to covering of the contaminated sediment with clean sediment. Thus, determining the distribution of contaminants in the bed sediments is important for delineating areas that need to be dredged for navigation and the subsequent mediation of the dredged contaminated sediments.

REFERENCES

Briggs, M. J., I. Melito, Z. Demirbilek, and F. Sargent. 2001. *Deep-draft entrance channels: Preliminary comparisons between field and laboratory measurements*. ERDC/CHL CHETN-IX-7. Vicksburg, MS: USACE, Research and Development Center.

CTH (Committee on Tidal Hydraulics). 1967. *Grays Harbor, Washington*. Vicksburg, MS: US Army Corps of Engineers.

CTH (Committee on Tidal Hydraulics). 1971. *Estuarine navigation projects*. T.B. 17. Vicksburg, MS: US Army Corps of Engineers.

EPA (US Environmental Protection Agency). 1992. *Evaluating environmental effects of dredged material management alternatives—A technical framework*. EPA842-B-92-008. Washington, DC: EPA and US Army Corps of Engineers.

Hayter, E. J., S. J. Smith, D. R. Michalsen, Z. Demirbilek, et al. 2012. *Dredged material placement site capacity analysis for navigation improvement project at Grays Harbor, WA*. ERDC/CHL TR-12-18. Vicksburg, MS: USACE, Research and Development Center.

Maynord, S. T., S. K. Knight, S. Bourne, M. R. Graves, et al. 2004. *Upper Mississippi River—Illinois Waterway System Models Report—Physical effects models*. Vicksburg, MS: USACE, Research and Development Center/Rock Island, IL: US Army Engineer District.

PIANC. 2023. *Beneficial use for sustainable waterborne transport infrastructure projects*. EnviCom Working Group Rep. No. 214. Brussels, Belgium: PIANC.

Pokrefke, T. J. 2012. *Inland navigation: Channel training works*. MOP 124. Reston, VA: ASCE.

USACE (US Army Corps of Engineers). 2002. *Coastal engineering manual*. Engineer Manual 1110-2-1100. Washington, DC: USACE.

USACE. 2015. *Dredging and dredged material management*. Engineer Manual 1110-2-5025. Washington, DC: USACE.

CHAPTER 4
SOLUTIONS TO SEDIMENTATION PROBLEMS

The variety of channel environments and sedimentation processes prevents a one-size-fits-all solution to deposition and erosion problems. A classification system—a taxonomy—and examples of solutions are presented subsequently and in Appendix B: Case Studies. They illustrate the variety of solutions that should be evaluated when choosing channel sedimentation solutions.

SOLUTION STRATEGIES AND METHODS

Solutions to sedimentation problems may take many forms, ranging from nonstructural to structural to dredging. Task Committees of the Coasts, Oceans, Ports, and Rivers Institute (COPRI), ASCE, and working groups of the World Association for Waterborne Transport Infrastructure (referred to as PIANC) formulated solution taxonomies to aid in solving navigation channel sedimentation issues (Pokrefke 2012, PIANC 2009). Those taxonomies are adapted here, as shown in Figure 4-1. The USACE's Regional Sediment Management approach employs many of these same strategies in a structured solution approach (USACE 2021c).

Briefly, Figure 4-1 consists of three categories—prevention, treatment, and adaptation—and seven strategies within those categories. Two strategies—keep sediment in place and dredge and place sediment—address excessive erosion. All categories and strategies address excessive deposition or turbidity. Note that some examples are encompassed in more than one strategy. For example, bypassing sediment to keep sediment moving around an inlet usually requires the dredge and place action to

Category	Strategy	Example Methods
Prevention	Keep Sediment in Place	Land erosion control Bed and bank erosion control
	Keep Sediment Out	Sediment traps Gates and dikes Channel separations Diversions
	Keep Sediment Moving	Training structures Agitation devices Flocculation reduction Bypass sediment
Treatment	Keep Sediment Navigable	Passive nautical depth Active nautical depth
	Dredge and Remove Sediment	Confined disposal Open water placement
	Dredge and Place Sediment	Beneficial uses Bypass sediment
Accommodation	Adapt to Sediment Regime	Water level timing Flexible infrastructure Offshore lightering Offshore terminals Passive nautical depth

Figure 4-1. Sedimentation solutions taxonomy.
Source: Adapted from Pokrefke (2012).

move it from the updrift side to the downdrift side of a coastal inlet. The application of various remedies can be seen in Appendix B.

The taxonomy in Figure 4-1 is recommended as a guide to choosing solutions for channel sedimentation problems. Considering a variety of solutions at the beginning allows the engineer to see beyond the standard narrow selection of either structures or dredging. Structures or dredging may be valid solutions, but the choice should be made in light of all options.

Most of the methods listed in Figure 4-1 will be familiar to sedimentation engineers. A few are discussed subsequently.

Keep Sediment in Place

Keeping sediment in place is a basic step in preventing excess deposition and turbidity. The Coastal Zone Management Reauthorization Act of 1990 required states to develop plans for controlling nonpoint source pollution, including land erosion. Federal guidelines (EPA 2021) and state guidance documents (e.g., ARC 2001) resulted from that requirement. Technical guidance for streambank erosion protection is provided by the Natural Resources Conservation Service (NRCS 1996) and the Corps of Engineers (e.g., Biedenharn et al. 1997; Watson et al. 1999; USACE 1994, 2002). The ASCE Sedimentation Manual (Garcia 2008) describes many erosion control methods.

Keep Sediment Out

Pascagoula Harbor, Mississippi, Navigation Project (Figure 4-2) provides an example of keeping sediment out. The project includes two sediment traps, one on the east side of the Pascagoula Channel that runs from the Gulf of Mexico to Horn Island Pass, and another trap on the east side of the channel through Horn Island Pass to the Gulf of Mexico.

The US Army Corps of Engineers (USACE) has studied constructing a sediment trap in the Head of Passes area (where the main Mississippi River channel divides into three distributary channels) to trap sediment in one portion of the river to minimize downstream deposition. (See the case study in Appendix B for maps.) The effects of Mississippi River diversions upriver of Head of Passes on shoaling in the navigation channel have also been studied; in the case of the mid-Barataria diversion, numerical models showed deposition just downstream of the diversion with no initial increase in dredging required to maintain authorized depth (USACE 2021a). Downriver of Venice, Lousiana, numerical modeling showed initially less dredging followed by a small increase 44 years after diversion begins. (Simulating multiple decades of sedimentation introduces additional uncertainty, as discussed in Chapter 5 and Appendix B.)

Some diversions in proximity to the navigation channel have had the opposite effect; a gap in the left descending bank of the lower Mississippi River near Buras has widened into an 850 ft wide channel that includes a 100 ft scour hole near its junction with the river. This diversion has resulted

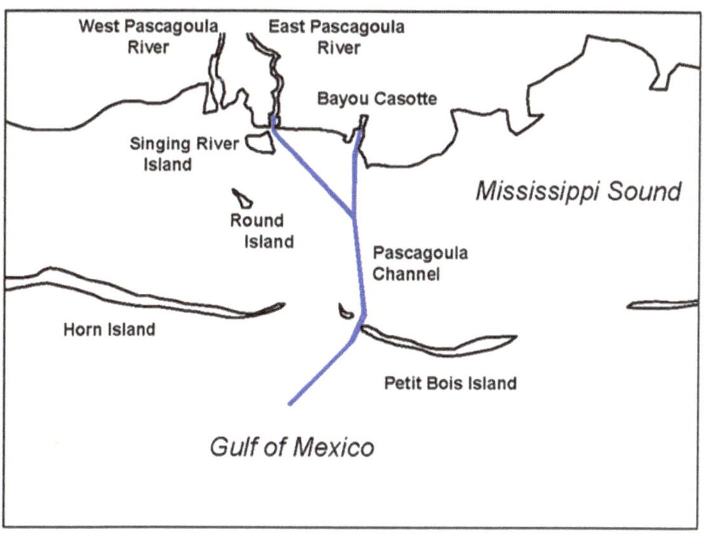

Figure 4-2. Pascagoula Harbor, MS, and Channels.
Source: Corson et al. (2003).

in dredging in a portion of the Mississippi River where there has never previously been shoaling (Schleifstein 2022). The diversion also has the potential, like Burrwood Bayou in Southwest Pass, to cause adverse current patterns in the navigation channel.

Keep Sediment Navigable

Keeping sediment navigable sounds like an oxymoron; however, Passive Nautical Depth is practiced worldwide in channels with fine-grained sediment through adoption of a nautical depth definition of the channel bottom based on a critical density of the sediment suspension (PIANC 2009). Acoustic fathometers show a false bottom at the depth of the lutocline, yet vessels safely sail through the high concentration, low-density suspension. Active Nautical Depth employs a specialized dredge that brings bed sediment to modified hoppers in which the sediment slurry slowly flows to overflow ports, becoming oxygenated. Upon its return to the water column, aerobic bacteria within the slurry release extracellular polymeric substances that hinder settling and consolidation of the suspension. Here, too, vessels can pass through the suspension using the nautical depth definition of the bottom (McAnally et al. 2007, PIANC 2009).

An example of the application of Active Nautical Depth is the port of Emden in Germany. Maintenance dredging averaged 4 million tonnes annually prior to the designation of the Nautical Depth; dredging was reduced to zero (Kirby 2015). Another example is the Atchafalaya Bar Channel, a portion of the Atchafalaya River and Bayous Chene, Boeuf, and Black navigation channels in the United States. This trench channel historically shoaled with rapidly consolidating fluid mud shortly after dredging completion. Funding was insufficient to maintain the channel year-round with conventional dredges. An offshore vessel was modified with jetted drag arms, a dredge pump, and piping to discharge material directly overboard. The vessel, shown in Figure 4-3, operates around the clock to keep fluid mud in the bar channel dispersed below 1.3 mg/L, a technique best described as agitation dredging to allow Passive Nautical Depth to be practiced.

Dredging

Three of the strategies in Figure 4-1 involve dredging. Dredge equipment can be divided into several categories. Mechanical dredges, operating from land or barges, remove material by excavation and include back-hoe, bucket ladder, clamshell bucket, and grab dredges. These dredges are used to remove hard-packed material or debris and can work in confined areas. Hydraulic dredges move sediments by pumping a water and sediment

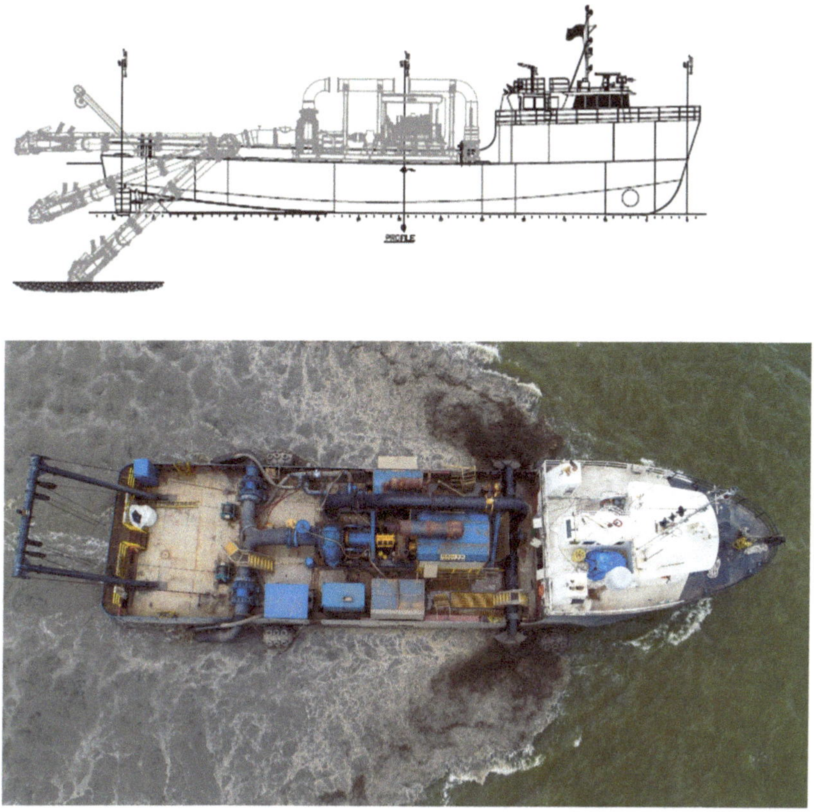

Figure 4-3. Dredge Arulaq, used to maintain the Atchafalaya Bar Channel, Louisiana. Top: Side view schematic showing dragarms rotation from horizontal to dredging position in contact with the bed. Bottom: Overhead photograph of dredge in operation with sediment discharges at surface on both sides.
Source: Courtesy of Brice Civil Constructors, Inc., a subsidiary of Calista Brice LLC. Used with permission.

slurry. Trailing suction hopper dredges are vessels that pump the sediment slurry, hold it internally in a hopper, and, once full, transport the dredged material to be disposed. A draghead is used to excavate the material and mix the material with water to create the slurry. Excess water is typically discharged as overflow. Once at the disposal site, material can be directly pumped out or dumped by split hulls or a bottom door. Hopper dredges can excavate sands, silts, clays, gravels, and some kinds of rock and can be used in shallow and deep channels. Because of their maneuverability, they can work in busy harbors and navigation channels.

Other types of dredges include specialized equipment such as water injection dredges, illustrated in Figure 4-4. Water injection dredges inject

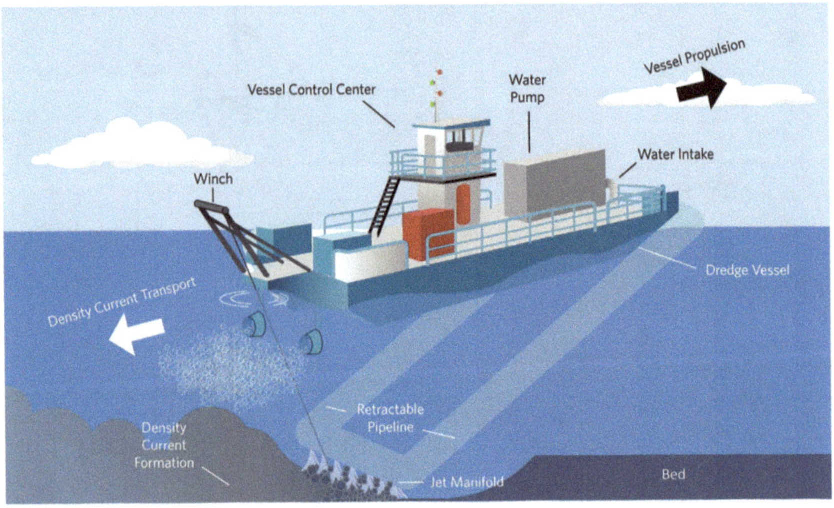

Figure 4-4. Schematic of a typical water injection dredge.
Source: Courtesy of HDR, Inc. Used with permission.

water into the channel bed, fluidizing the sediment so that it can flow to deeper areas without the need for transportation to a disposal site. The approach is suited for locations with silts, unconsolidated clays, and fine sands and nearby areas to which the fluidized sediment can flow without problems. It is often used in smaller ports for maintenance dredging. The dredge can operate near structures and moored vessels. Projects in Wilmington and Morehead Harbors, North Carolina, and three locations in the Mississippi River have demonstrated the water injection dredging method (Fuller et al. 2023).

After material is dredged by traditional methods, it must be disposed. There are several disposal strategies: open water disposal, sustainable relocation, beneficial use, confined disposal, and treatment. Open water disposal can also be conducted at designated sites in the ocean, estuaries, rivers, and lakes. Open water disposal can be dispersive or retentive. For some dredging projects, such as the Mississippi River crossings, the material is dispersed into the river downstream of the crossing. For retentive disposal, material may also be capped to isolate it from the environment. Sustainable relocation is commonly employed for offshore sediments; dredged sediments are disposed back into the littoral currents.

A beneficial use of sediments applies them productively, such as for coastal protection, land creation or reclamation, habitat enhancement, and construction material (PIANC 2023a). For example, dredged material from the Atchafalaya River and Bayou Chene, Boeuf, and Black navigation channel has been used to enhance delta growth and create bird islands

(Foran et al. 2018). For the Mississippi River deep-draft project, 49 mi^2 of land was created between 1975 and 2014 using between 30% and 50% of available and suitable dredged sediments from Head of Passes and Southwest Pass channels (USACE 2012).

With confined disposal, dredged material is placed inside dikes or other constructed structures. The water within the dredged material is allowed to run off or evaporate. Liners (compacted fine sediment, synthetic membranes, or grout mattresses) may be present, and the water may be treated before being discharged.

Further information on dredging equipment and methods can be found in Randall (2022), the USACE (2015) manual *Dredging and Dredged Material Management* and on the web at Central Dredging Association (2005), and at web sources such as Western Dredging Association (WEDA 2021) and the International Association of Dredging Companies (IADC 2021).

Adapt to Sediment Regime

Water-level timing uses high waters, typically from tides, to provide depths sufficient to navigation. Examples include half-tide harbors in Alaska and elsewhere that go dry during the low water tidal phase and fill with water during the high water phase, allowing temporary access. The Port of Hamburg, Germany, employs the approach for a shallow zone in the Elbe River about halfway between the port and the North Sea. Vessels drafting above a critical value enter the Elbe and leave the port on schedules that place them at the shallow zone during high tide.

SUSTAINABLE REMEDIES

> Sustainable water resource systems are those designed and managed to meet the needs of people living in the future as well as those of those of us living today. (ASCE 1998)

Evolving Environmental Concerns

Most early navigation projects were constructed and maintained with minimal regard for depletion or destruction of natural resources. Projects were often of too modest a scale for large environmental impacts and science was inadequate to define all but the most obvious effects. Design, construction, and operation of early ports and waterways (e.g., the ports at Alexandria, Egypt, about 330 BCE and St. Augustine, Florida, about 1565 CE) took advantage of water naturally deep enough to pass the ships of the day and employed at most small structures to provide safe anchorage. Whatever impact they had on their surroundings was ignored unless the port itself became threatened. Resources seemed abundant in relation to their consumption.

Over the next few centuries, the growing importance of commercial navigation forced engineers to consider approaches employing new knowledge of the physical sciences in designing and maintaining navigation channels. In the 1850s, Marie Joseph Raymond Thomassy proposed to the New Orleans Drainage Commissions the building of a number of sediment diversions out of the Mississippi River to reduce navigation channel sedimentation and provide flood protection. This same trend, now called "building with nature" or "engineering with nature" occurred in Europe and Asia in the 19th century (Hooimeijer and Lambert 2021).

The dawn of the modern environmental consciousness in the United States is often linked to two events—the 1962 publication of Rachel Carson's *Silent Spring*, and the first Earth Day in 1970. Carson's book forecasted the loss of thousands of wild species and harm to humans if widespread indiscriminate use of pesticides continued unabated. Despite furious criticism, accumulating evidence proved Carson right. In 1969, Senator Gaylord Nelson called for a nationwide grassroots demonstration on behalf of the environment on April 22, 1970. He invited everyone to participate, and more than 20 million people responded. Despite opposition to the concept of environmental protection from some industries and governmental skepticism about the threat of human activities to the environment, public opinion grew more and more supportive of environmental protection. Congress began passing laws designed to reduce environmental damage, including the Water Quality Act of 1965, National Environmental Policy Act of 1969, and Clean Water Act of 1972, which contributed to slowing or halting the environmental degradation Carson and Nelson had warned of.

The earliest US environmental law specifically related to navigation—the 1899 US Rivers and Harbors Act—was designed to protect navigation. The Act banned dumping of refuse or fill into waterways to protect shipping. Only in the 1960s was the prohibition on dumping of "refuse" begun to be interpreted as preventing pollution.

Environmental laws and regulations gradually changed the practice of water resources engineering. At first, designers worked primarily to ensure that no laws or regulations were violated while designing for function and cost efficiency, as evidenced by design guidance wording such as "fully complying with all applicable environmental laws." Gradually, the design emphasis changed to reflect avoidance of environmental harm, with guidance using phrases such as, "… minimize or eliminate adverse effects to the environment …" (USACE 1980) More recently, the emphasis has evolved toward natural resources stewardship and the goal of sustainable development. For example, the Tennessee Valley Authority has adopted as policy, "... promote proactive environmental sustainability in a balanced and ecologically sound manner," although it still retains the older mindset with a policy of, "... maintaining compliance with all applicable environmental and legal requirements" (TVA 2021).

Sustainability

In addition to laws, regulations, and policy, the principles of sustainable development have been incorporated into codes of ethics. For example, the ASCE code of ethics (ASCE 2020) includes the following provision:

> [Engineers shall]
> a. adhere to the principles of sustainable development;
> b. consider and balance societal, environmental, and economic impacts, along with opportunities for improvement, in their work;
> c. mitigate adverse societal, environmental, and economic effects; and
> d. use resources wisely while minimizing resource depletion.

Despite this and similar statements in other engineering codes of ethics, sustainability remains almost as controversial as basic environmental protection was in the 1970s. A frequent criticism is that true sustainability is idealistic and impossible—any use of resources is bound to decrease the amount available to future generations. However, that criticism is no more valid than saying that we need not design and build for safety, because perfect safety is never achieved. Absolute sustainability can be an ideal goal that is balanced against other goals, like economic development, and sustainability can become a design criterion, properly defined.

The ASCE sustainability definition given at the start of this article echoes ASCE Policy 418 (ASCE 2021):

> The American Society of Civil Engineers (ASCE) defines sustainability as a set of economic, environmental, and social conditions (aka "The Triple Bottom Line") in which all of society has the capacity and opportunity to maintain and improve its quality of life indefinitely without degrading the quantity, quality, or the availability of economic, environmental, and social resources. Sustainable development is the application of these resources to enhance the safety, welfare, and quality of life for all of society.

Waterborne transportation is widely understood to be the most economical form of transport. Less well known is that it may have less impact on the air and water quality and natural habitat than highway, railway, pipeline, or air transport. For example, inland water transport of freight consumes much less petroleum products (thus producing lower emissions) per ton-mile of transport than highway transport and somewhat less than rail transport (Casavant 2000). Although reliable comparative metrics for other forms of resource consumption and degradation are unavailable, waterborne transport may be also more environmentally friendly in terms of habitat fragmentation, water pollution, and habitat destruction. Responsible decision-making on transportation policy and investment requires true multimodal performance metrics, including those for sustainability. An ASCE/UNESCO project has offered suggestions for measuring sustainability (ASCE 1998) that can be combined with intermodal transportation metrics (e.g., McAnally et al. 2004) to support decision-making in transportation.

Definitions and policies on sustainability have been adopted by multiple organizations. Some, such as PIANC (2003, 2023b, c), have translated sustainability policies into recommended design processes, and others, such as USACE (2021d), have expressed them in manuals and regulations. PIANC (2003) recommended that navigation project analyses account for key physical and ecological processes, including

- Morphologic processes,
- Hydrologic processes,
- Sedimentation balance,
- Habitat provision, and
- Biological and chemical processes.

In addition to a recommended analysis procedure, the cited PIANC (2003) report offers design alternatives that should be considered, such as ensuring adequate waterway depth by adapting vessel loads, training works, and flow regulation, as well as lock and dam structures and dredging.

PIANC (2023b) provides inland waterway managers with a guide to the processes for more sustainable contributions to society as a whole through waterways' multiple uses. Cases studies describe projects that successfully implemented sustainable solutions.

Sustainability Rating Systems

Evaluating the relative sustainability of solutions is not a casual endeavor; for example, the Envision rating system has more than 60 scoring criteria in five categories (ISI 2021):

- Quality of life (14 criteria),
- Leadership (12 criteria),
- Resource allocation (14 criteria),
- Natural world (14 criteria), and
- Climate and risk (10 criteria).

One project related to a navigable waterway was identified as an Envision award winner. The "Life MarinaPlan Plus" project at the Marina of Cervia, Italy, on the Adriatic Sea won the Institute for Sustainable Infrastructure's Silver Award for "Promotion of a more sustainable development of economic activities in marine environment, ports in particular" in 2021 (ISI 2022). The system consists of 10 jet pumps and collectors that fluidize sediment deposits on the updrift side of the marina entrance and pipelines that transport the sediment–water mixture to the downdrift side, as shown in Figure 4-5. The system provides a full-depth entrance channel with reduced costs and reduced environmental impact (Pellegrini et al. 2021).

Sediment bypassing projects similar to the Cervia system have a long record of reducing or eliminating dredging while keeping sediment moving at coastal inlets in Australia and the United States (Clausner 1990, 2000).

Applying sustainability criteria is site and project-specific. Nevertheless, the strategies of Figure 4-1 can be evaluated for some relative measures of sustainability. For example, maintenance dredging that must be regularly repeated is inherently less sustainable than a one-time fix of keeping sediment out by preventing erosion. Similarly, adopting a Nautical Depth

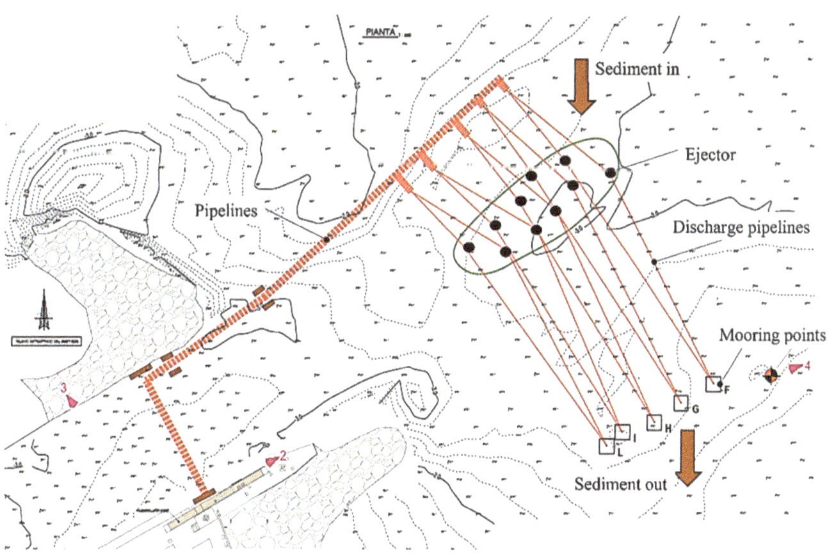

Figure 4-5. Sustainability award-winning sediment bypassing system at Cervina, Italy.
Source: Adapted from original provided by Prof. Marco Pellegrini. Used with permission.

definition of the bottom is more sustainable than repeatedly dredging large volumes of muddy water (PIANC 2009)

The strategies of Figure 4-1 are listed subsequently in approximately descending order of sustainability, recognizing that individual applications of each strategy may not fall in this order and that reasonable people may disagree.

- Adapt to sediment regime,
- Keep sediment in place,
- Keep sediment out,
- Keep sediment navigable,
- Keep sediment moving,
- Dredge and remove sediment, and
- Dredge and place sediment.

Holism

The sustainability concept has expanded to include holistic solutions. Smuts (1926) is credited with coining the English term *holism*, which he described as "the tendency in nature to form wholes that are greater than the sum of the parts through creative evolution." Smuts saw holism as an active force in the universe, responsible for organizing "wholes." He defined wholes as "... composites which have an internal structure, function, or character which clearly differentiates them from mere mechanical additions or constructions... ." He described as wholes a water molecule (more than a simple mixture of hydrogen and oxygen atoms), cells (more than a collection of water, minerals, and organic molecules), organism (more than a collection of cells), and the universe. We might add ecosystems, societies, and waterborne transportation systems to his list of wholes.

The holism definition has been refined and applied in diverse fields, most vividly by Douglas Adams (1987) as the "fundamental interconnectedness of all things." Adams' definition helps to remind us first, that economic development and a healthy ecosystem are fundamentally connected as interacting contributions to the quality of life, and second, that what happens in a navigation channel affects other, often unseen aspects and areas of the human ecosystem.

Holism in sedimentation solutions has been applied in a number of organized endeavors, including

- Regional sediment management (RSM). "A systems approach using best management practices for more efficient and effective use of sediments in coastal, estuarine, and inland environments = Healthy Systems" (USACE 2021c).

- Engineering with nature (EWN). "EWN is the intentional alignment of natural and engineering processes to efficiently and sustainably deliver economic, environmental, and social benefits through collaborative processes" (USACE 2021b).
- Building with nature (BwN). "Building with Nature (BwN) is a design philosophy in which natural processes deliver a number of benefits such as protection against flooding and coastal erosion, and opportunities for nature and recreation" (EU 2021).
- Envision. "... provides a flexible system of criteria and performance objectives to aid decision makers and help project teams identify sustainable, resilient, and equitable approaches during the planning, design, and construction that will continue throughout the project's operations, maintenance, and end-of-life phases" (ISI 2021).

Channel sedimentation predictions are commonly used to anticipate the sedimentation effects of a range of solutions, as addressed previously in this chapter. Some of the remedy examples in Figure 4-1 fit sustainability and holistic concepts better than others.

Sustainable or holistic solutions should be part of a thorough sedimentation solution evaluation. For example, agitation dredging might be a cost-effective and efficient way to solve a channel deposition problem, but sustainability criteria require that annualized costs, equipment availability, and equipment engine fuels and emissions be evaluated along with the sediment removal efficiency. A holistic evaluation would further expand the required considerations to local and downstream habitat and socioeconomic effects. Although standard sedimentation prediction methods may not be designed to address these nonphysical consequences, it may be necessary to choose solution methods or add analyses that will supply information to evaluations by interdisciplinary teams.

Loucks (2003) and Stakhiv (2011) have suggested that managing water resources from a holistic perspective include social justice or environmental justice, defined by USACE (2022) as, "fair treatment and meaningful involvement of all people regardless of race, color, national origin, or income regarding the development, implementation, and enforcement of environmental laws, regulations, and policies, with no group bearing a disproportionate burden of environmental harms and risks." Incorporating environmental justice into a project requires effective communication with all communities affected by the project and collaboration across the professional stovepipes (Fields et al. 2020).

REFERENCES

Adams, D. 1987. *Dirk Gently's holistic detective agency*. New York: Pocket Books.

ARC (Atlanta Regional Commission). 2001. *Georgia stormwater management manual, Vol. 2: Technical handbook*. Atlanta: ARC.

ASCE. 1998. *Sustainability criteria for water resource systems*. ASCE Task Committee on Sustainability Criteria. New York: ASCE.

ASCE. 2020. *Code of ethics*. Reston, VA: ASCE.

ASCE. 2021. *Policy 418*. Reston, VA: ASCE.

Biedenharn, D. S., C. M. Elliott, and C. C. Watson. 1997. *The WES stream investigation and streambank stabilization handbook*. Vicksburg, MS: USACE Waterways Experiment Station.

Casavant, K. 2000. *Inland waterborne transport—An industry under siege*. Washington, DC: US Dept. of Agriculture, Agricultural Marketing Service.

Central Dredging Association. 2005. "Dredging: The facts." Accessed January 12, 2021. https://dredging.org/media/ceda/org/documents/resources/cedavarious/publications-dredging_the_facts.pdf.

Clausner, J. E. 1990. *Fixed sand bypassing plant—An update*. DRP-3-03, Dredging Research—Technical Notes. Vicksburg, MS: USACE, Research and Development Center.

Clausner, J. E. 2000. Sand bypassing cost and performance database. ERDC/CHL TN-II-41. Vicksburg, MS: USACE, Research and Development Center.

Corson, W. D., B. A. Ebersole, L. Lin, D. J. Mark, et al. 2003. *Evaluation of island and nearshore confined disposal facility alternatives, Pascagoula River Harbor dredged material management plan*. ERDC TR-03-3. Vicksburg, MS: USACE, Research and Development Center.

EPA (US Environmental Protection Agency). 2021. *Nonpoint source pollution: Technical guidance and tools*. Washington, DC: EPA.

EU (European Union). 2021. "Building with nature." Interreg, North Sea Region. Accessed June 5, 2021. https://building-with-nature.eu.

Fields, N. L., V. J. Miller, C. Cronley, K. K. Hyun, et al. 2020. "Interprofessional collaboration to promote transportation equity for environmental justice populations: A mixed methods study of civil engineers, transportation planners, and social workers' perspectives." *Transp. Res. Interdiscip. Perspect*. 5 (2020): 100110.

Foran, C., K. Burks-Copes, J. Berkowitz, J. Corbino, et al. 2018. "Quantifying wildlife and navigation benefits of a dredging beneficial-use project in the lower Atchafalaya River: A demonstration of engineering with nature®." *Integr. Environ. Assess. Manage*. 14 (6): 759–768.

Fuller, W. P., R. J. Wagner, and R. E. Lewis. 2023. "U.S. dredging challenges and opportunities—Hydrodynamic dredging." In *Proc., Transportation Research Board 102nd Annual Meeting Conf*. Washington, DC.

Garcia, M. 2008. *Sedimentation engineering: Processes, measurements, modeling, and practice*, MOP 110. Reston, VA: ASCE.

Hooimeijer, F. L., and D. Lambert. 2021. "Building with nature, a nineteenth century concept." *J. Delta Urbanism* 2 (2021): 24–47. https://doi.org/10.48438/jdu.2.2021.6223.

IADC (International Association of Dredging Companies). 2021. "Equipment." Accessed October 2, 2023. https://www.iadc-dredging.com/tag/equipment-publication.

ISI (Institute for Sustainable Infrastructure). 2021. "Envision." Accessed May 21, 2021. http://www.sustainableinfrastructure.org.

ISI. 2022. "Marina plan plus project." Accessed January 12, 2021. http://www.sustainableinfrastructure.org.

Kirby, R. 2015. "Module 6 nautical depth." *Presentation to Ports and Maritime Organization of 3rd Int. Iran and PIANC Iran Conference*, Teheran.

Loucks, D. P. 2003. "Managing America's rivers: Who's doing it?" *Int. J. River Basin Manage.* 1 (1): 21–31.

McAnally, W. H., C. Friedrichs, D. Hamilton, E. Hayter, et al. 2007. "Management of fluid mud in estuaries, bays, and lakes, part 2: Measurement, modeling, and management." *J. Hydraul. Eng.* 133 (1): 23–38.

McAnally, W. H., Y. Zhang, A. J. Allen, R. O. Bowden, et al. 2004. *Transportation responses to increased Latin American trade.* Denver: National Center for Intermodal Transportation, University of Denver and Mississippi State University.

NRCS (Natural Resources Conservation Service). 1996. *Streambank and shoreline protection, chapter 16 of engineering field handbook.* Washington, DC: USDA.

Pellegrini, M. G., B. Prada, M. Mikac, C. Ponti, et al. 2021. "Life Marina Plan Plus project: Sustainable marine and coastal seabed management." *Terra et Aqua.* No. 162, Spring.

PIANC (World Association for Waterborne Transport Infrastructure). 2003. *Guidelines for sustainable inland waterways and navigation.* Report of Working Group 6 of the Environmental Commission. Brussels, Belgium: PIANC.

PIANC. 2009. *Minimising siltation in harbours.* PIANC Rep. 102. Brussels: PIANC.

PIANC. 2023a. *Beneficial use for sustainable waterborne transport infrastructure projects.* EnviCom Working Group Rep. No. 214. Brussels: PIANC.

PIANC. 2023b. *Guidelines for sustainable recreational navigation infrastructure.* RecCom Working Group Rep. No. 148. Brussels: PIANC.

PIANC. 2023c. *Sustainable inland waterways: A guide for inland waterway managers on social and environmental impacts.* InCom Working Group Rep. 203. Brussels, Belgium: PIANC.

Pokrefke, T. J. 2012. *Inland navigation: Channel training works.* MOP 124. Reston, VA: ASCE.

Randall, R. E. 2022. *Dredging and dredged material placement*. Reston, VA: ASCE.

Schleifstein, M. 2022. "Power of the river: New Mississippi River channel prompts Corps to take action." *The Times-Picayune New Orleans Advocate*.

Smuts, J. C. 1926. *Holism and evolution*. Sherman Oaks, CA: Reprinted in 1999 by Sierra Sunrise Publishing.

Stakhiv, E. Z. 2011. "Pragmatic approaches for water management under climate change uncertainty." *J. Am. Water Resour. Assoc.* 47 (6): 1183–1196.

TVA (Tennessee Valley Authority). 2021. *Environmental policies and principles*. Knoxville, TN: TVA.

USACE (US Army Corps of Engineers). 1980. *Layout and design of shallow-draft waterways*. EM 1110-2-1611. Washington, DC: USACE.

USACE. 1994. *Hydraulic design of flood control channels*. EM 1110-2-1601, Revised. Washington, DC: USACE.

USACE. 2002. *Coastal engineering manual*. Engineer Manual 1110-2-1100. Washington, DC: USACE.

USACE. 2012. *Mississippi river navigation and marsh restoration*. Washington, DC: USACE.

USACE. 2015. *Dredging and dredged material management*. Engineer Manual 1110-2-5025. Washington, DC: USACE.

USACE. 2021a. *Draft environmental impact statement for the proposed mid-Barataria sediment diversion project, Plaquemines Parish, Louisiana*. Appendix Q. New Orleans: USACE.

USACE. 2021b. *Engineering with nature*. Vicksburg, MS: USACE. Accessed May 21, 2021. https://ewn.el.erdc.dren.mil/.

USACE. 2021c. *Regional sediment management*. Vicksburg, MS: USACE. Accessed May 21, 2021. https://rsm.usace.army.mil/.

USACE. 2021d. *Sustainability*. Washington, DC: USACE. Accessed December 21, 2020. https://www.usace.army.mil/Missions/Sustainability.aspx.

USACE. 2022. *Environmental justice*. Washington, DC: USACE. Accessed August 27, 2023. https://www.usace.army.mil/Missions/Environmental-Justice/.

Watson, C. C., D. S. Biedenharn, and C. R. Thorne. 1999. *Demonstration erosion control design manual*. Vicksburg, MS: USACE, Research and Development Center.

WEDA (Western Dredging Association). 2021. "Western Dredging Association home page." https://www.westerndredging.org/.

CHAPTER 5
PREDICTION METHODS

Channel sedimentation responses to the solutions described in Chapter 4 can be predicted by numerous methods grouped into four general approaches:

- Field observations,
- Desktop analyses,
- Physical models, and
- Numerical models.

These approaches, described as follows, vary in terms of data requirements, complexity, reliability, applicability, cost, and time requirements. They should be applied with full recognition of their strengths and weaknesses and the user's skill in applying them.

FIELD OBSERVATIONS

Field observations provide insight into the physical processes of sedimentation and data for desktop analyses, physical modeling, and numerical modeling. Basic data collected for sedimentation studies include standard time-varying morphologic, hydrodynamic, sediment, and water quality measurements:

- Bathymetric/hydrographic surveys and topographic surveys.
- Hydrodynamics
  - Water levels, including hydrographs, tides, and surges, as appropriate;
  - Upstream inflows;

  - Meteorology;
  - Vertical and horizontal flow-velocity profiles; and
  - Wave field characteristics.
- All sediment
  - Grain size distribution in bed, banks, and inflows;
  - Organic content;
  - Sediment inflow rates;
  - Vertical and horizontal suspended sediment profiles;
  - Sediment transport rates;
  - Rates of deposition and erosion;
  - Dredging records.
- Noncohesive sediment: sediment bed-density profiles and sediment bed characteristics (e.g., size distribution, profiles).
- Cohesive sediment
  - Mineral grain size;
  - Settling velocity distributions;
  - Sediment bed-density profiles;
  - Sediment bed-rheologic characteristics;
  - Depositional and erosional characteristics.
- Water quality
  - Temperature;
  - Salinity;
  - Organic content; and
  - Sediment contaminant profiles.

Other information may be required by some analyses and specialized instruments (e.g., densimeters, pore pressure gauges). Methods for noncohesive sediment data collection are described in Chapter 5 of Garcia (2008) and cohesive sediment data collection are described by Mehta (2018) and Hayter et al. (2014). Bruner et al. (2020) describe field data requirements for hydrodynamic modeling, many of which are also needed for sedimentation studies.

The Federal Interagency Sedimentation Project (FISP) was created in 1939 to standardize fluvial sediment research and development activities across federal agencies. Today, seven US federal agencies are involved: Department of Agriculture, US Army Corps of Engineers, US Geological Survey, Bureau of Land Management, Bureau of Reclamation, Environmental Protection Agency, and Forest Service. Federal Interagency Sedimentation Project (FISP) instrumentation and research addresses a broad range of topics related to the measurement and analysis of suspended sediment, bedload sediment, bed material, bed topography, and water quality. Examples of FISP work include field testing of acoustic backscatter (ABS) sediment sensors, bedload monitoring using hydrophones, in situ bedload measurements using multiple acoustic methods, pressure-difference bedload samplers, and in situ densimetric measurements as a surrogate for suspended sediment concentrations. Additional information on FISP can be found at the FISP web site.

Sediment

Bed material gradation data provide key input for sedimentation studies. USGS (2007) provides guidance on sampling bed material in rivers and streams. As noted in the manual, sampling should be sufficient to determine the spatial variability, size, and gradation of the bed material. Field conditions may dictate the type of sampler and procedure used.

In coastal and estuarine environments, bottom sediment samples can be collected on a regular grid and also used to identify likely net sediment pathways and regions of erosion, accretion, and dynamic equilibrium (Hughes 2005).

The amount of sediment suspended in the water column is quantified by a variety of terms and measurement techniques that are usually incompatible and frequently confusing. Commonly used terms include

- Suspended sediment concentration (SSC),
- Total suspended solids (TSS), and
- Turbidity.

Although SSC and TSS are defined as the mass of material with size greater than 2 μm in a given volume of water, they are measured by different procedures. Standard lab procedures for TSS measurement involve extracting a subsample of water for filtration, drying, and weighing (APHA 1995). The lab procedure for SSC requires use of the entire water sample (ASTM 2000). The difference in measurement techniques can produce substantially different results, and US Geological Survey (USGS) prescribes SSC as the only appropriate method to quantify suspended sediment in US waters (Gray et al. 2000). Note that some documents and data sources, particularly older ones, do not make this distinction and use the terms TSS or SSC for suspended sediment measured or calculated by alternate techniques, such as turbidity.

Turbidity is a measure of water clarity and can be affected by both dissolved and suspended materials. For example, decaying vegetation may leach dissolved tannins into waterbodies, dyeing the waters so that they are quite turbid but may contain little sediment. Similarly, turbidity measurements are also sensitive to the particle size of the suspended sediments—per unit mass, silts, and clays will include more particles than sand-sized particles and thus have more impact on turbidity.

Turbidity measurement data in the literature include Secchi Disk readings, Nephelometric Turbidity Units (NTU), Formazin Turbidity Units (FTU), and even the older Jackson Turbidity Units. Optical backscatter devices measure NTU by light scattered from a sample. Turbidimeters measure light attenuation through a sample and display either FTU or NTU values.

Although widely used in water-quality work, turbidity measurements must be used with caution in sedimentation studies because no constant, reliable conversion exists between turbidity and TSS or SSC. The USGS provides guidelines for computing time-series SSC from turbidity sensor data that can be used by experienced users (Rasmussen et al. 2011); however, the results may be limited to the site, the season, and sometimes even the time of day (Pratt and Parchure 2003). See the section on *Error and Uncertainty in Field Observations* for additional information.

Similar to SSC estimated from turbidity, SSC profiles can also be estimated from ABS measured by acoustic Doppler current profilers (ADCPs). The measured ABS profiles can be used with ABS-SSC correlations (developed using SSC measurements in water samples) to develop estimates of SSC profiles with depth in the water column and over time (in case of a fixed long-term ADCP deployment) or instantaneous across the flow cross section (in case of boat-based ADCP transects). In combination with velocity measurements from the ADCP, the ABS-estimated SSC can also be used to estimate instantaneous SSC fluxes and correlated to environmental forcings relevant for sediment transport (Wall et al. 2006, Scully and Friedrichs 2007, Mathew and Winterwerp 2020). However, ABS measurements are sensitive to the shape and size of the suspended sediments, and the resulting SSC estimates should be used with caution in sedimentation studies.

Bedload measurement technology consists primarily of pressure-difference samplers and surrogates. This technology has been used mostly for streams with gravel size or larger bed material. Surrogates include acoustic doppler bottom tracking, passive acoustics, seismometers, and impact sensors.

An emerging technology in bedload measurement is the ISSDOTv2 method that uses time-sequenced bathymetric data of bed forms to determine a bedload transport rate (Abraham et al. 2019). This method has been validated with flume data; it provides consistent and repeatable bedload measurements in riverine environments where the predominant bedload transport mode is bed forms. It has been used as part of a study to forecast Chippewa River sediment entering the Mississippi River, which causes channel closures and navigation delays.

Deposition and Erosion

Rates of channel deposition and erosion can be estimated by hydrographic/bathymetric surveys analysis, densiometric surveys analysis, dredging records, sediment sample dating, and test pits.

Hydrographic Surveys. Bathymetry is a critical component of any assessment of navigation channels and the projection of sedimentation rates. Applications of bathymetric assessment include the determination of current conditions of the channel and assessment of historical and future sedimentation rates. The primary tool in the measurement of bathymetry is hydrographic surveying.

Standards for Hydrographic Surveys (USACE 2013) provide the US industry standards for hydrographic surveys of navigation, water control, and shore protection projects. Technical guidance in the manual covers a wide range of surveying techniques and the information necessary to specify, conduct, and interpret hydrographic surveying information. The manual standards are used for the vast majority of surveys in US waters. Additional international standards are available from the International Hydrographic Organization in Special Publication No. 44 (IHO 2020). Standards of the IHO may be required for coastal projects, projects for private entities, and projects extending into international waters; however, surveys meeting USACE standards will also typically meet IHO standards. Any surveys used in the assessment of navigation channels must meet these widely accepted guidance and standards documents. Although this section will cover the most commonly used techniques for surveying, many other technologies are available beyond the scope of this document and are covered in more detail in USACE (2013).

Hydrographic survey technology today is primarily based on acoustic measurements. An acoustic transducer system deployed from a vessel transmits an acoustic beam(s) that bounces off the bed and the reflected signal(s) is recorded by the system. The surveying system controlling the transducer measures the two-way travel time for an individual acoustic pulse (or ping) between transducer and receiver. Using the position of the transducer, typically determined by global positioning system on the vessel, and the travel time of the acoustic ping through the water, the depth to the bed can be determined for that position. Higher acoustic frequencies are typically used (> 100 kHz) to reflect from the initial sediment water interface at the bed. Lower frequencies (< 50 kHz) may also be used in specific application to penetrate through lower-density material at the bed surface. However, reflected acoustic signals are generated by a density gradient and cannot detect a specified density.

The standard technology for many surveys is a single-beam sonar. A vertical beam is transmitted from a single transducer mounted to a vessel. The position and depth at the vessel are recorded from each ping as the vessel transits survey lines. To convert the depth to a datum-referenced elevation, the water surface elevation is measured with a locally surveyed gauge. Single-beam surveys are often performed by the vessel transiting cross sections across the channel at regular distance intervals (e.g., every 100 ft). Overall, single-beam sonar is a low-cost system often used because of the ease of deployment, cost-effectiveness, and rapid data processing. The major drawback of single-beam surveying is the lack of continuous coverage of the seabed. Measurements are available only directly below the lines of the vessel transits.

A more comprehensive survey coverage is possible with a multibeam sonar. Multibeam uses a swath of acoustic beams across the bottom, which can be configured to cover up to 180 degrees on either side of the vessel. The broader coverage allows for overlapping sweeps of the bottom during vessel transits. Because of the multiple beams, multibeam can provide one hundred percent coverage of the bed, allowing for continuous bathymetric maps. Multibeam sonar is more expensive and requires more processing time for the data; however, the costs and time are often outweighed by the significant increase in survey coverage. Multibeam is the most widely used technology in coastal regions and is commonly used in all navigation channels.

Dual beam sonar is often used to investigate the potential thickness of lower consolidated, or sometimes referred to as unconsolidated, sediment at the surface of the bed. A lower-frequency ping 10 to 50 kHz penetrates the less-consolidated surface with a higher frequency ping of > 100 kHz used to reflect from the consolidated sediment surface. Although the technique is subjective for defining sediment layers, it can be useful in evaluating sediment conditions at some sites where large quantities of lower consolidation state sediment are present (USACE 2013). There are serious challenges with quantitative interpretation in areas with strong suspension lutoclines, fluid mud, or highly mobile sediment beds, and densiometric surveys are recommended instead of acoustics (McAnally et al. 2007).

Aircraft LiDAR (light detection and ranging) scanning systems such as the USACE Coastal Zone Mapping and Imaging system (Tuell et al. 2010, Irish et al. 2016) can accurately map bottom depths, provided turbidity is not excessive. For example, Feygels et al. (2012) reported a successful mapping of depths of 7 to 8 m on the Mississippi Gulf of Mexico coast when the Secchi attenuation coefficient was 0.46 m^{-1}. The USACE Joint Airborne LiDAR Technical Center of Expertise (JALBTCX 2023) offers a web repository of survey data and tools for public use.

For navigation channel sedimentation assessment, a comparison of hydrographic surveys is one of the most common ways to estimate historical rates of sedimentation. Although it is best to define survey specifications a priori for comparative surveys, historical surveys may be the best available. Identifying survey specifications according to the USACE (2013) guidance is required to best define sources of uncertainty in each individual survey so that any errors in comparison can be defined. Examples of uncertainty also include the challenges with comparing single-beam surveys where survey lines do not overlap and multibeam surveys using different equipment and calibration standards. Further, incorporation and interpolation of hydrographic data in geographic information systems (GIS) systems may be done differently introducing further errors and uncertainty. Although a comparison of two or more hydrographic surveys remains a primary method for estimating historical sedimentation, it is critical that any assessment fully considers all sources of error and uncertainty. The USACE CSAT computer program provides tools for computing changes between hydrographic surveys (Dunkin et al. 2018)

US Army Corps of Engineers has an online navigation portal in which hydrographic surveys and channel boundaries, centerlines, and stationing for USACE maintained federal navigation channels are accessible (see https://navigation.usace.army.mil/Survey).

Digital elevation models (DEM) store, display, and manipulate elevation data—topographic, hydrographic, and bathymetric information. Typically used in GIS, they can also constitute one of the tool sets provided in a numerical model program's associated graphical user interface (GUI), which are described in the *Numerical Models* section.

Dredging Records. Past dredging volumes, identified by time period and channel reach, provide useful data for validating prediction methods and establishing rough relationships between deposition volumes and environmental events such as storms and floods. However, annual or seasonal dredged volumes may have been constrained by funding, equipment availability, weather, or environmental windows and not reflect the actual accumulated sediment volume for that period. More useful are pre-dredge and post-dredge hydrographic surveys of channels, which can be analyzed for depositional and erosional volumes between dredging events.

USACE District offices maintain dredge records for the federal navigation channels for which the agency is responsible. Ports also may keep dredging records for areas of their responsibility.

Test Pits. Digging test pits and measuring sediment accumulation in them provides an estimate of the deposition rate that may be expected if sediment supply and hydrodynamics are unchanged. However, projects such as channel enlargement or realignment often alter the hydrodynamic and sedimentation regimes, and therefore, the method must be employed with care. Mikkelsen et al. (1980) described a field experiment in which 200 m wide, 400 m long, and 2 m deep test pit was dredged in the Warri Port Channel of Nigeria and measured pit deposition by echosounder every three weeks for 5 months. Those and other data were analyzed to determine the dominant transport mechanisms and test a sedimentation prediction method.

Sediment Dating. Radionuclide dating methods for measuring sedimentation rates involve the measurement of radioactive isotope activity over depth in sediment cores (Holmes 1998). These methods can also be used for dating specific sediment strata. Typical isotopes used are Cesium-137 (Cs137) and Lead-210 (Pb210) (Erickson et al. 2007). Cesium-137, with a half-life of 30.3 years, is associated with the atmospheric fallout of the by-products of thermonuclear testing which started in about 1954 and peaked in 1963; atmospheric fallout of Cs137 has been declining since 1963. These time horizons are typically apparent in the measured Cs137 activity in individual core profiles, especially in consistently depositional settings, and can be used to develop an estimate of the long-term sedimentation rate for individual cores. In contrast, Pb210, with a half-life of 22.3 years, is a naturally occurring isotope of atmospheric origin, and delivered to the aquatic environment by precipitation, followed by adsorption to depositing sediment. This produces a profile of decreasing Pb210 activity with depth in the sediment bed that, given the known half-life of Pb210, can be used to estimate the sedimentation rate. However, because both methods rely on measurements on sediment cores, they can be spatially limited. In addition, estimated sedimentation rates are typically most reliable in consistently depositional settings rather than settings subject to episodic erosion and deposition or otherwise disturbed by processes such as dredging and vessel propwash and reflect the net sedimentation rate over long periods of time rather than short-term variations. Therefore, the estimated sedimentation rates should be carefully interpreted and considered with other lines of evidence.

Tracing Sediment

The simplest sediment tracing techniques consist of in situ sediment grains, painted or dyed for later identification, placed in erosional areas, and then sampled after transport to determine direction and quantities of movement. Distinctively dyed grains filling a trench also provide a direct measure of erosion depth. Such color-tagged sediment can be hard to locate in vigorous transport environments if they are widely dispersed and buried by untagged sediment.

Tagging sediment grains with radioisotopes such as Scandium-46 was once thought to be a promising tracing technique. Minute quantities of the tagged sediment could be detected in sediment beds, even if covered by nontagged sediment; however, the difficulties and dangers of working with even mildly radioactive materials seriously limited its use. Isotopes already in the environment, including both those naturally occurring, such as Pb210, and those added by weapons testing and nuclear power generation, such as Cesium-137 and Chromium-51, respectively, can be used to date deposits or trace sediment paths (e.g., Corcoran and Kelley 2006, Robertson and Fix 1977).

Naturally occurring characteristics, such as minerology, distinctive grain shape, magnetic properties, and the aforementioned natural isotopes, have all been employed to identify sediment sources and paths with varying levels of success. Walling (2013) summarizes many of these methods.

Other Data

If available, historical sediment observations provide insight into how sediment entering the study reach has changed over time. Historical charts and maps, from which shoreline and bankline changes are measured, can be used to estimate a part of the sediment supply. For instance, the National Oceanic and Atmospheric Administration (NOAA 2022) maintains a publicly accessible database of historical nautical charts from the 1800s onward, which can be used to develop a conceptual understanding of morphological changes, shoreline evolution, infrastructure development, and so on. Specific gauge analyses of water-level data can be used to detect aggradation and degradation trends in the study reach. Bank and shoreline erosion rates measured with erosion pins can be useful for the identification of sediment sources (e.g., Ramirez-Avila et al. 2010).

Error and Uncertainty in Field Observations

Field observations contain calibration errors (including meter drift and fouling), postprocessing calculation errors, and spatial and temporal errors. The latter errors affect water and sediment discharge observations most particularly because calculations require assumptions about spatial and temporal integration. Examples of observed sediment data errors are plentiful. Optical sensors readily measure turbidity, but there is no physical correlation between turbidity and SSC. General trends can be established, but changes in sediment color (e.g., phytoplankton under sunny versus cloudy conditions), presence of silica, and size distribution can alter the calibration (Sutherland et al. 2000). There has been success using high-quality optical backscatter measurements in conjunction with the same site, same day calibrations to SSC measurements, but the calibrations may not suffice for other sites or the same site on a different day if conditions change. Extreme care is required (Downing 2006, Pratt and Parchure 2003, Rasmussen et al. 2011).

Hydrographic surveys over large areas have been successfully used to identify areas of deposition and erosion; however, horizontal uncertainties ranging from 1 to 5 m and vertical uncertainties of 0.1 to 2 m are possible, and qualified surveyors using the best available standards should be consulted in any assessment (USACE 2013, IHO 2020). These uncertainties arise from many factors, including reference benchmark; reference water-level gauges and extrapolation of water levels across the surveyed reach, geoid, tidal phase and range; survey boat static and dynamic draft; survey boat stability and orientation; depth measurement system; and bottom material density variation. USACE (2013) describes these uncertainties and provides examples of navigation project depth measurement uncertainties.

Dredging volumes are subject to the uncertainties and accuracies of the surveys from which they are calculated. Hourly dredge production rates may provide insight into dredging volumes but are subject to dredge factors (e.g., pump size, speed of vessel), environmental factors (e.g., sediment characteristics, wind, depth), and operational factors (e.g., crew efficiency). Dredged volumes may not correspond to sediment deposition volumes because of funding or schedule constraints on dredging.

DESKTOP METHODS

Desktop approaches—using equations and principles that can be solved without automated numerical methods—represent the simplest methods for estimating sedimentation rates and can be applied relatively efficiently. Most desktop methods are empirical in nature, relying on field observations to predict future sedimentation rates. Their benefits of speed and low cost arise from assumed simplifications to physics and geometry and those same simplifications often produce lesser reliability than more rigorous approaches. Desktop analyses for the Houston Ship Channel, Norfolk Harbor, Tenn-Tom Waterway, and Atchafalaya Bay are described in Appendix B.

Computers have blurred the lines between desktop methods and numerical modeling. Calculations that were once performed with pencil, paper, and calculator (and slide rules before that) can be automated in spreadsheets and scripts in MATLAB, R, and other utilities. We begin with obvious pencil and paper methods and progress to machine learning on small computers that are called models.

General Desktop Approaches

Trapping Efficiency. Lake and reservoir sedimentation studies sometimes employ simple trapping calculations to estimate loss of storage capacity from sediment accumulation. The technique does not directly apply to navigation channel deposition but can be used as part of a sediment budget or analyzed to extract a rough estimate of channel deposition in a navigation pool (e.g., see the Tenn–Tom case in Appendix B).

The trapped sediment amount is calculated as $S = Eff * SY$, where S is the trapped sediment amount, Eff is the fractional trapping efficiency, and SY is the sediment yield of the watershed draining to the pool. Trapping

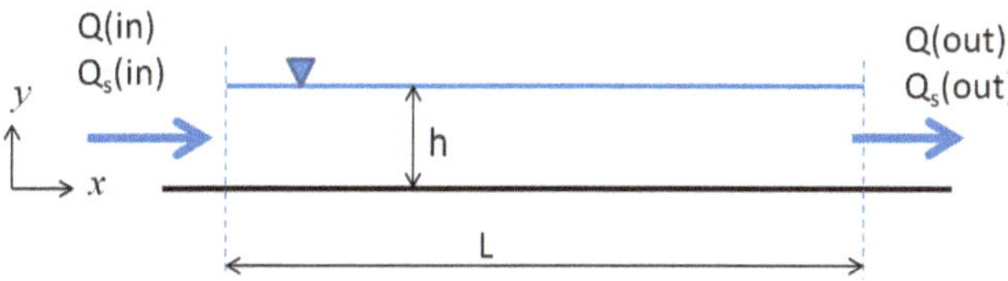

Figure 5-1. Flow profile schematic.

efficiency is a function of inflowing sediment concentration and settling velocity, navigation pool storage capacity relative to inflow rate, pool shape, and dam operation. In practice, efficiencies are taken from empirical curves, some of which are presented by USACE (1989).

A more rigorous trapping efficiency is based on a general numerical solution for transport, settling, deposition, and erosion of sediment grains in a rectangular channel, as shown schematically in Figure 5-1, where Q is the water discharge, Q_s is the sediment discharge, L is the reach length, and h is the flow depth.

This method does not require that the reach be controlled by a dam at the lower end but, like the pool trapping efficiency, it assumes that deposition occurs uniformly throughout the reach. It can be used for reach deposition estimation or sediment trap design.

For transport and settling of incoming sediment load, Q_s, the three-dimensional (3D) advection–diffusion equation (Garcia 2008) is reduced to the simplified, steady-state two-dimensional (2D) form:

$$u\frac{\partial c}{\partial x} - W_s\frac{\partial c}{\partial y} = \frac{\partial}{\partial x}\left(E_y\frac{\partial c}{\partial y}\right) \tag{5-1}$$

where

u = Flow speed in the x-direction (horizontal streamwise),
c = Sediment concentration (variable in x and y),
W_s = Sediment settling velocity, and
E_y = Eddy diffusion coefficient in the y-direction (vertical), which can be estimated by the expression

$$E_y = 0.75\sqrt{g}\,\frac{nhU}{K_m R^{1/6}} \tag{5-2}$$

where

g = Acceleration of gravity,
n = Manning roughness coefficient,

h = Water depth,
U = Average horizontal flow speed,
K_m = Units conversion factor (1 in SI units, 1.49 in customary units), and
R = Hydraulic radius (cross-sectional area divided by wetted perimeter).

These equations assume that

- Lateral (bank to bank) flow variation can be represented by an average value at each location in the vertical dimension.
- Turbulent diffusion of sediment concentration is equal to turbulent diffusion of water mass and is not affected by sediment in suspension.
- Longitudinal diffusion can be neglected.

Equation (5-1) can be solved numerically, subject to boundary conditions of zero flux at the water surface and zero concentration gradient at the bottom plus the following assumptions:

- Concentration profile at the upstream boundary is constant.
- Vertical velocity profile is logarithmic.
- Vertical eddy diffusion coefficient profile is parabolic.
- Sediment particles that touch the bed are deposited.

Generalized numerical solutions have been produced by a number of authors, notably by Sarikaya (1977), in which the trapping efficiency is expressed by the nondimensional equation

$$Eff = \frac{Q_s(\text{in}) - Q_s(\text{out})}{Q_s(\text{in})} = F\left\{7.5\frac{W_s}{u_*}, \frac{W_s L}{Uh}\right\} \qquad (5\text{-}3)$$

where

$Q_s(\text{in})$ = Inflowing sediment load,
$Q_s(\text{out})$ = Outflowing sediment load,
$F\{\}$ = Function of the variables inside the brackets,
W_s = Sediment settling velocity,
L = Length of reach,
U = Depth-averaged flow speed,
h = Depth of reach, and
u_* = Shear velocity, given by Equation (2-1).

A solution set of these equations is displayed in Table 5-1. Knowledge of the variables in Equation (5-3) can be used to pick off values of the trapping efficiency ranging from 0.006 (0.6%) to 1 (100%). The table shows that for increasing settling velocity and length of impoundment and decreasing shear velocity, flow speed and depth, the retention ratio increases.

Table 5-1. Solutions for Trapping Efficiency as Expressed by Equation (5-3).

	W_sL/Uh																
$7.5W_s/u_*$	0.01	0.1	0.2	0.3	0.4	0.5	0.6	0.7	0.8	0.9	1	1.1	1.2	1.5	2	5	10
0.001	0.006	0.08	0.16	0.23	0.28	0.32	0.32	0.32	0.32	0.30	0.31	0.32	0.33	0.42	0.65	0.74	0.83
0.01	0.007	0.08	0.16	0.24	0.29	0.35	0.37	0.39	0.41	0.41	0.43	0.46	0.47	0.56	0.73	0.79	0.85
0.05	0.008	0.09	0.16	0.24	0.31	0.36	0.41	0.44	0.47	0.49	0.52	0.55	0.57	0.65	0.79	0.83	0.87
0.1	0.008	0.08	0.16	0.23	0.29	0.35	0.41	0.46	0.51	0.54	0.58	0.62	0.65	0.73	0.83	0.87	0.93
0.2	0.008	0.08	0.16	0.24	0.30	0.36	0.41	0.46	0.51	0.55	0.60	0.63	0.66	0.74	0.84	0.89	0.94
0.3	0.008	0.09	0.16	0.24	0.31	0.37	0.43	0.47	0.53	0.57	0.61	0.64	0.67	0.76	0.85	0.90	0.94
0.4	0.009	0.09	0.16	0.25	0.31	0.38	0.44	0.48	0.54	0.58	0.62	0.65	0.68	0.77	0.86	0.90	0.95
0.5	0.009	0.09	0.17	0.25	0.32	0.39	0.44	0.49	0.54	0.59	0.63	0.66	0.69	0.78	0.87	0.91	0.96
0.6	0.009	0.09	0.17	0.25	0.32	0.39	0.45	0.50	0.55	0.59	0.64	0.67	0.70	0.78	0.87	0.91	0.96
0.7	0.009	0.09	0.17	0.25	0.33	0.40	0.45	0.51	0.56	0.60	0.64	0.67	0.71	0.79	0.88	0.92	0.97
0.8	0.009	0.09	0.17	0.25	0.33	0.40	0.46	0.52	0.57	0.61	0.65	0.68	0.72	0.80	0.88	0.92	0.97

(Continued)

Table 5-1. (*Continued*).

$7.5W_s/u_*$	W_sL/Uh 0.01	0.1	0.2	0.3	0.4	0.5	0.6	0.7	0.8	0.9	1	1.1	1.2	1.5	2	5	10
0.9	0.009	0.09	0.17	0.25	0.33	0.40	0.46	0.52	0.57	0.62	0.66	0.69	0.73	0.81	0.89	0.93	0.97
1	0.009	0.09	0.17	0.26	0.34	0.41	0.47	0.53	0.58	0.63	0.66	0.70	0.74	0.82	0.90	0.94	0.98
2	0.009	0.09	0.17	0.26	0.34	0.43	0.50	0.56	0.62	0.67	0.71	0.76	0.79	0.86	0.94	1.00	1.00
3	0.010	0.09	0.17	0.26	0.35	0.43	0.51	0.58	0.65	0.70	0.75	0.79	0.83	0.90	1.00	1.00	1.00
4	0.010	0.09	0.17	0.26	0.35	0.44	0.52	0.59	0.66	0.73	0.77	0.82	0.85	0.92	1.00	1.00	1.00
5	0.010	0.09	0.17	0.26	0.35	0.44	0.52	0.60	0.67	0.74	0.78	0.83	0.86	0.93	1.00	1.00	1.00
6	0.010	0.09	0.17	0.26	0.35	0.44	0.52	0.61	0.68	0.75	0.79	0.84	0.88	0.94	1.00	1.00	1.00
7	0.010	0.09	0.17	0.26	0.35	0.44	0.53	0.62	0.69	0.76	0.81	0.85	0.89	0.95	1.00	1.00	1.00
8	0.010	0.09	0.17	0.26	0.35	0.44	0.53	0.62	0.70	0.76	0.82	0.86	0.90	0.96	1.00	1.00	1.00
9	0.010	0.09	0.17	0.26	0.35	0.44	0.53	0.63	0.70	0.77	0.82	0.87	0.91	1.00	1.00	1.00	1.00
10	0.010	0.09	0.17	0.26	0.35	0.44	0.53	0.63	0.71	0.77	0.83	0.88	0.91	1.00	1.00	1.00	1.00

(*Continued*)

Table 5-1. (*Continued*).

	W_sL/Uh																
$7.5W_s/u_*$	0.01	0.1	0.2	0.3	0.4	0.5	0.6	0.7	0.8	0.9	1	1.1	1.2	1.5	2	5	10
20	0.011	0.09	0.17	0.26	0.35	0.44	0.54	0.63	0.72	0.80	0.86	0.91	0.94	1.00	1.00	1.00	1.00
30	0.011	0.09	0.17	0.26	0.35	0.44	0.54	0.64	0.73	0.81	0.87	0.92	0.95	1.00	1.00	1.00	1.00
40	0.011	0.09	0.17	0.26	0.35	0.44	0.54	0.64	0.73	0.82	0.88	0.93	0.96	1.00	1.00	1.00	1.00
50	0.011	0.09	0.17	0.26	0.35	0.44	0.54	0.64	0.74	0.83	0.89	0.94	0.97	1.00	1.00	1.00	1.00
60	0.011	0.09	0.17	0.26	0.35	0.44	0.54	0.64	0.74	0.83	0.89	0.94	0.97	1.00	1.00	1.00	1.00
70	0.011	0.09	0.17	0.26	0.35	0.44	0.54	0.64	0.74	0.83	0.90	0.94	1.00	1.00	1.00	1.00	1.00
80	0.011	0.09	0.17	0.26	0.35	0.44	0.54	0.64	0.74	0.83	0.90	0.95	1.00	1.00	1.00	1.00	1.00
90	0.011	0.09	0.17	0.26	0.35	0.44	0.54	0.64	0.74	0.84	0.90	0.95	1.00	1.00	1.00	1.00	1.00
100	0.011	0.09	0.17	0.26	0.35	0.44	0.54	0.64	0.74	0.84	0.91	0.96	1.00	1.00	1.00	1.00	1.00

Bottom Area Method. The *bottom area*, or *footprint* method for predicting deposition rates is not recommended. It assumes that the sedimentation rate per unit area will be the same in a modified channel as in the existing channel, so the predicted sedimentation volume is calculated as the product of the existing sedimentation rate and the modified channel bottom area (typically widening or lengthening the channel). The basic assumption—that hydrodynamics and sediment transport processes will remain constant when a channel is enlarged—is obviously wrong, yet the method was once popular and still occasionally appears in technical reports. Trawle and Herbich (1980) found that the bottom area method grossly miscalculated the actual sedimentation volume in coastal and estuarine channels. They found that the *volume-of-cut* method to be a more accurate predictor of actual sedimentation volumes than other desktop methods.

Volume-of-Cut Method. In the volume-of-cut method (Trawle and Herbich 1980, Trawle 1981, Van Rijn 2013), the volume of sediment deposited in the channel is related to the volume of sediment that was removed beyond the natural depth or the depth of the surrounding waterway (i.e., the volume of cut) to create a channel of given dimensions or to expand an existing channel. Accordingly, this is expressed as follows:

$$V_{\mathrm{Dep}} = \gamma V_{\mathrm{Cut}} \tag{5-4}$$

where

V_{Dep} = Volume of sediment deposited in the channel on an annual basis,
V_{Cut} = Volume of cut, and
γ = Proportionality factor.

For the case of the deepening of an existing navigation channel, γ may be calculated using the known maintenance dredging rates and the existing V_{Cut}. The sedimentation rate to be expected in the future can subsequently be calculated by applying γ to the future V_{Cut}, that is, after deepening.

If channel enlargement changes hydrodynamics by too much, methods such as volume of cut may give erroneous results, typically underestimating the enlarged channel sedimentation rates. For example, if an estuarine channel experiences significant differences in salinity intrusion as a result of channel deepening, the locations and magnitudes of sedimentation often change dramatically (e.g., Bruun 1988, CTH 1995). The volume-of-cut method is best used to give an order of magnitude estimate of increased sedimentation based on the enlargement of a channel cross section. These estimates can be used in conceptual studies of alternatives.

Inland Channel Methods

Sharp (2007) developed a three-tiered sediment budget method to define sediment inflows, outflows, and net deposition rates for Aberdeen Pool on the Tenn–Tom Waterway. By progressing from a simple, rapid method to more rigorous methods, he showed that the user could cross-check results and gain an improved understanding of the sedimentary system. The method consists of three steps and produced deposition estimates as described in the Tenn–Tom section of Appendix B.

The HEC-RAS modeling system (see subsequent *Numerical Modeling* section) contains several modules that are useful for initial sediment assessments. The Bank Stability and Toe Erosion Model (BSTEM), developed by the National Sediment Laboratory, US Department of Agriculture, provides two spreadsheet components: geotechnical bank failure using two methods and toe scour erosion that computes bank undercutting by hydraulic forces. The results from the BSTEM module can be integrated back into an HEC-RAS movable-bed sediment model. Midgley et al. (2012) applied it to Barren Fork Creek in Oklahoma and found that it accurately predicted the timing of bank failures but slightly underpredicted the rate of streambank retreat in that application. It can be used to estimate the contribution of bank failure to channel deposition.

SIAM is a sediment budget tool within HEC-RAS that can be used to compare relative channel responses to a range of alternatives. The module has six transport functions that compute annualized transport capacity by grain size. Output can be the annualized surplus or deficit for a reach, sediment transport potential, or supply and balance plots divided into bed supply and wash supply components.

HEC-RAS also includes channel design functions. Stable channel dimensions are computed using three different methods. Six sediment transport functions are available to calculate sediment transport capacity.

Vessel-induced wave suspension of near-channel sediments and their subsequent channel deposition can be calculated by means of desktop methods. Parchure et al. (2001) used basic mathematical expressions and a simplified approach to estimate suspension of both cohesive and noncohesive sediments by vessel-generated waves. Their simplified approach, implemented in a desktop computer program, provided preliminary estimates of suspension concentration magnitudes and subsequent deposition for the Upper Mississippi River channel.

Coastal Channel Methods

Some coastal desktop approaches include a representation of the physical processes responsible for sedimentation. These are based on simplified theoretical considerations, including the relationship between

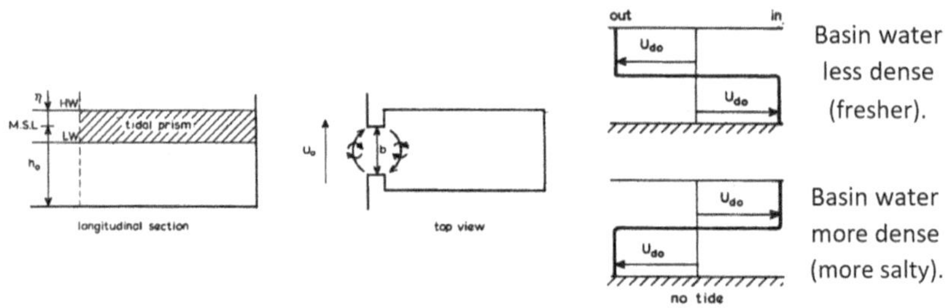

Figure 5-2. Schematic of volume exchange for tidal basins: tidal exchange (left), horizontal eddy filling (middle), and density-driven circulation (right). Source: Adapted from Eysink and Vermaas (1983).

the hydraulics, physical dimensions of the channel/basin, sediment characteristics, and the process of sedimentation. Because these approaches include an analytical representation of the physical processes involved, they may be considered more reliable and appropriate than strictly empirical methods for estimating the impact of a change in the forcing functions responsible for transport and sedimentation.

One such analytical desktop approach, applicable to sedimentation in harbor basins, involves the formulation of Eysink and Vermaas (1983). Figure 5-2 shows a schematic of the hydraulics represented in the formulation—these include the volume exchanges associated with (1) the barotropic tide (V_t), (2) horizontal eddy exchange at the mouth of the basin (V_h), and (3) baroclinic circulation (V_d). Accordingly, the total hydraulic volume exchange per tidal cycle (V_e) is

$$V_e = V_t + V_h + V_d \tag{5-5}$$

The inflowing water delivers suspended sediments to the basin, of which a fraction p (also referred to as the sediment trapping efficiency) is deposited in the basin as a function of the sediment settling velocity, water depth, hydraulic retention time, and flow velocity. The sedimentation rate is calculated as follows:

$$\Delta T_S = \frac{p\bar{c}_{\text{in}} V_e}{\rho_d} \tag{5-6}$$

where

ΔT_S = Sedimentation volume per tide,
$\bar{c}_{\text{in}}$ = Tidal mean suspended sediment concentration of the inflowing water, and
ρ_d = Dry bulk density of the sediment deposits.

Eysink and Vermaas (1983) include formulations for estimating terms V_t, V_h, V_d, and p as a function of various physical parameters such as the salinity gradient, velocity gradient, water depth, sediment settling velocity, and so on. An application of the method is described in the Norfolk Harbor section of Appendix B.

Another analytical desktop approach, applicable to sedimentation in a channel, trench or pit, involves the formulation of Van Rijn implemented as a spreadsheet tool called SED-PIT (Van Rijn 2013). This method is applicable to navigation channels cutting through coastal zones and wide basins. Briefly, this approach involves the calculation of the change in flow velocity within the channel (owing to the larger water depth relative to areas outside the channel), and therefore, the reduction in sediment transport capacity within the channel. The plan view on the upper panel of Figure 5-3 shows a schematic of flow (from the bottom left to the upper right of the panel) over a channel oriented at an arbitrary angle relative to the direction of tidal currents. The reduction in flow velocity within the channel is calculated using basic equations of motion and continuity. In addition to a reduction in velocity within the channel, the currents also experience a change in direction. The velocity component perpendicular to the channel experiences a reduction in magnitude that is inversely proportional to the local water depth, whereas the velocity component parallel to the channel may increase in magnitude because of a reduction of bottom friction. As a result, the streamlines show a refraction-type pattern in the channel (Figure 5-3). Associated with suspended sediment

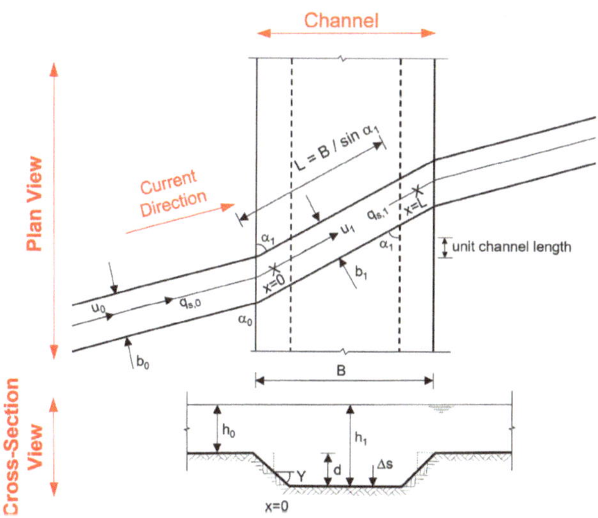

Figure 5-3. Schematic of flow over a channel oriented at an arbitrary angle relative to the tidal current direction.
Source: Adapted from Van Rijn (2013). Used with permission.

concentration, the sediment load entering the channel is calculated, and using the trapping efficiency of the channel (a function of the flow velocity, settling velocity, and channel geometry), the sediment deposition rate is calculated. In addition to deposition associated with advection (i.e., transport of suspended sediment with the tidal currents), wave-driven sediment mobilization from the seafloor adjacent to the channel and potential deposition within the channel can also be calculated.

Accordingly, the instantaneous sedimentation volume (ΔS) is calculated as follows:

$$\Delta S = \frac{\left[\left(\frac{b_0}{b_1}\right) q_{S,0} - q_{S,1}\right] e_S L \sin \alpha_1}{\rho_d} \tag{5-7}$$

where

b_0 and b_1 = Streamtube width outside and inside the channel,
$q_{S,0}$ and $q_{S,1}$ = Transport at the upstream edge of the channel and within the channel and calculated using the equilibrium transport formulation of Van Rijn (2013),
e_S = Sediment trapping efficiency, and
L and α_1 = Channel geometry and orientation as shown in Figure 5-3.

The calculations can be performed using time-variable tidal currents and wave conditions, or probabilistically given the annual wave climate for an estimate of the long-term average sedimentation rate.

The desktop approaches described previously were applied in the Norfolk Harbor case study (Appendix B) for various sections of the harbor and approach channels as appropriate given the physical setting. In addition, the volume-of-cut method served as an additional quantitative estimate of sedimentation rates for comparison to estimates from the approach of van Rijn (2013).

The coastal "Channel Shoaling Analysis Tool" (CSAT) (Dunkin et al. 2018) forecasts future dredging requirements of fixed navigation channels by using historical channel surveys and a hindcasting algorithm, based on past shoaling rates and location patterns, including inputs from longshore or cross-shore sediment movement. It can incorporate additional new hydrographic surveys to update the prediction of future volumes of sediment required for dredging at various time and depth intervals. It has been specifically created to take advantage of standardization in the uploading of channel hydrographic surveys and processing of data through the eHydro program (USACE 2022). The CSAT was used for the development of shoaling rates for the Houston Ship Channel study (Appendix B).

A coastal sediment budget defines sources and sinks of littoral, estuarine, and riverine sediment within a specific coastal region, and the likely linkages of sediment movement within the region. Sediment budgets can be developed for historical and existing conditions, and a predictive or theoretical sediment budget can be derived for future conditions on the basis of modification of features within the region (additional material placed on updrift shorelines, channel deepening, new structures that impact sediment movement such as groins, jetties, and breakwaters). Detailed historic data are needed on shoreline positions (erosion and accretion patterns), ebb and flood shoal growth patterns, channel dredging histories, and estimations of onshore/offshore movement of sediment. Sediment budgets can be used as analytical tools to predict and define navigation channel sedimentation patterns, especially in cases where there may be impacts to the channel owing to large updrift beachfill projects.

The Sediment Budget Analysis System (SBAS) software for ArcGIS©10 toolbar (Dopsovic et al. 2002) provides a convenient graphical interface for constructing coastal sediment budgets. Figure 5-4 shows a typical output for the New Jersey coast.

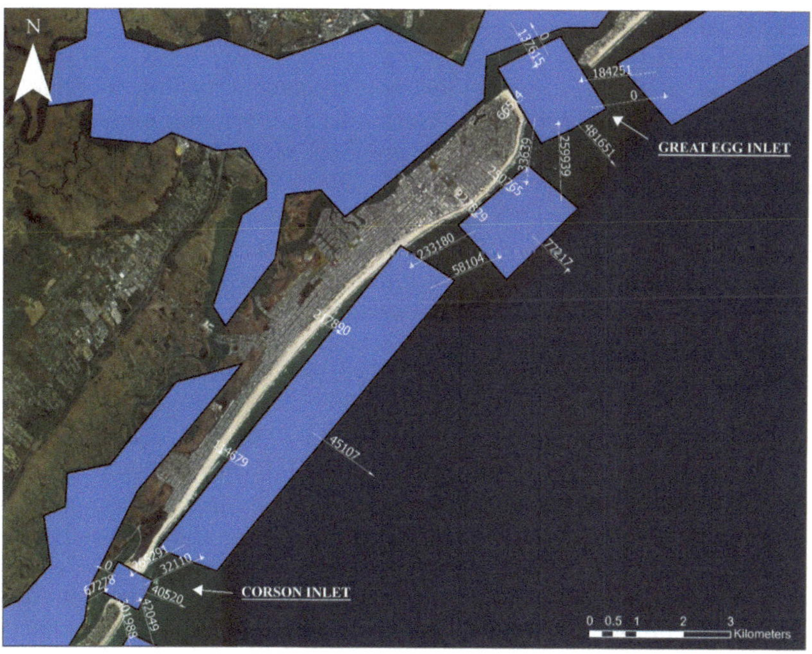

Figure 5-4. SBAS example sediment budget output for Ocean City, New Jersey, spanning from Corson Inlet to Great Egg Inlet, volumes in m³/year.
Source: Courtesy of USACE.

Machine Learning Models

Machine learning models use sophisticated regression techniques such as artificial neural network (ANN), fuzzy logic, and decision trees (e.g., Bhattacharya et al. 2007, Prasadreddy and Visweswararao 2019). Of these, only ANN has been commonly used in water resources modeling and most often employed to predict flows and sediment discharges instead of channel sedimentation. In that usage, it could support sediment budget calculations, discussed previously, and supply boundary conditions for the other methods described here. See Gupta et al. (2021) for examples of that use.

At the time of this manual's publication, artificial intelligence—use of computer hardware and software to perform tasks usually requiring human intelligence—has become newsworthy for applications used in many new fields, including some related to modeling. As with any new technology, applications of artificial intelligence to sedimentation issues should be carried out with extreme care, using the peer-reviewed literature and experienced sedimentation engineers as guides to appropriate practices and results.

PHYSICAL MODELS

Physical models—scale models constructed according to scientific scaling principles for geometric, kinematic, and dynamic similarity—are described in virtually all hydraulic engineering textbooks, as they have been used for several hundred years to solve open channel flow problems. They have also been long used for sedimentation problems, although sediment scaling relationships are less well defined.

Physical models are designed by enforcing the scaling of one or a series of dimensionless relationships between the model and the prototype (nature). It is not possible to scale all relationships, so the dominant ones are selected, and the remaining force relationships provide constraints for the model dimensions and characteristics.

Physical modeling requires considerable expertise, typically gained through long experience. Further, physical modeling of sediment transport is part science and part art because of the difficulty in developing scaling relationships that are useful and productive (Letter and McAnally 1977, Maynord 2006). For these reasons, most sediment transport modeling is now performed using numerical modeling, as discussed in the following section. An understanding of physical modeling is important to the sedimentation engineer, because the literature contains a wealth of prior physical model studies that can inform future work.

Model Scales

Open channel flows are most often modeled according to Froude scaling, in which the Froude number, the ratio of inertial and gravity effects given by Equation (5-8a), is equal in the model and prototype. Three major, but not scaled exactly, relationships are the constraints that the Reynolds number [Equation (5-8b)] exceeds the laminar threshold (guaranteeing that the model flow regime is turbulent), the wall Reynolds number or particle Reynolds number exceeds the near-bed laminar threshold (guaranteeing that the respective boundary layer is fully turbulent), and the Weber number [Equation (5-8c)] is large enough (greater than order 10, guaranteeing that surface tension effects are not a significant force in the model). In all three equations, V is the characteristic velocity, h is the flow depth or other characteristic length, g is the acceleration of gravity, ν is the kinematic viscosity, and σ is the surface tension. Given a length scale, $h_r = h_p/h_m$, where the subscripts p and m represent prototype (the original, or natural system) and model, respectively; the velocity scale ratio is determined for constant g. An example set of scale ratios for hydraulic processes is shown in Table 5-2.

$$F = V/\sqrt{gh} \tag{5-8a}$$

$$R = \frac{Vh}{\nu} \tag{5-8b}$$

$$W = \frac{\rho V^2 h}{\sigma} \tag{5-8c}$$

For environments dominated by long waves (length greater than 20 times the water depth), such as riverine, lacustrine and tidal flows, model horizontal and vertical length scales are usually different (a distorted scale model) to ensure fully turbulent model flows with Reynolds Number above the turbulent threshold. A secondary benefit of scale distortion is reduction of space requirements. Short wave (length less than two times the depth)-dominated environments such as coastlines and some bays and lakes are often modeled with undistorted length ratios or small distortion ratios. In cases where both long and short waves are significant factors, compromises in scaling must be made.

A variety of scaling relationships have been employed in modeling noncohesive sediments. The most common approach to developing scaling criteria for noncohesive sediment transport is to use forms of dimensionless

Table 5-2. Scale Ratios for a 1:100 Vertical and 1:500 Horizontal Scale Physical Hydraulic Model.

Parameter	Scale ratio, model/prototype	Value, model/prototype	Comment
Vertical length	$h_r = \frac{h_m}{h_p}$	$\frac{1}{100}$	Chosen to best represent phenomena of interest, keep flow turbulent, and minimize surface tension effects
Horizontal length	$L_r = \frac{L_m}{L_p}$	$\frac{1}{500}$	Chosen to best represent phenomena of interest, minimize space requirements, and meet surface roughness criteria
Length scale distortion	$N_r = \frac{h_r}{L_r}$	$\frac{5}{1}$	
Gravity	$g_r = \frac{g_m}{g_p}$	1	Unless built in a centrifuge for increased gravity or not on earth
Froude number	$F_r = \frac{V_r}{\sqrt{g_r h_r}}$	1	Specified as 1 to satisfy Froude similarity
Horizontal velocity	$V_r = \sqrt{g_r h_r}$	$\frac{1}{10}$	From Froude similarity
Time	$t_r = \frac{L_r}{\sqrt{h_r}}$	$\frac{1}{50}$	From velocity similarity
Water kinematic viscosity, surface tension, and density	$\nu_r = \sigma_r = \rho_r$	1	Same fluid as prototype
Reynolds number	$R_r = \frac{V_r h_r}{\nu_r} = \frac{h_r^{3/2}}{\nu_r}$	$\frac{1}{1000}$	Flow velocities large enough to maintain R_r greater than about 2,000
Weber number	$W_r = \frac{\rho_r V_r^2 h_r}{\sigma_r} = h_r^2$	$\frac{1}{10,000}$	Depths great enough to keep W_r greater than about 10

Shields parameters—a particle densimetric Froude number and a particle Reynolds number—given by

$$F_* = \frac{u_*^2 \rho}{\rho' g d} \tag{5-9}$$

$$R_* = \frac{u_* d}{\nu} \tag{5-10}$$

where

ρ = water density,
$\rho' = \rho_s - \rho$,
d = representative grain size,
ν = kinematic viscosity of water, and
ρ_s = sediment grain density.

Although it is commonly cited, Shields scaling is strictly a scaling of the initiation of motion and does not scale sediment transport. Pugh (2007) provides an extensive examination of sedimentation scaling relationships.

An additional constraint that arises with mobile bed physical models is the cohesive limit. If the prototype sediment is noncohesive, the model grain size cannot be finer than the cohesive limit (approximately 62 μm). This constraint is often addressed by utilizing materials in the model with a lower specific gravity than the prototype material. Lower specific gravities are also limiting in that fine, light-weight materials tend to float because of surface tension and adhering tiny air bubbles. Some examples of this approach are discussed in the next section.

Wave models in which both refraction and diffraction are significant processes are often constructed with undistorted scales—the vertical and horizontal scales are equal—to accurately model both processes. In a distorted scale model, either diffraction or refraction must be chosen to be modeled precisely, and the wave period ratio then chosen accordingly. Depending on the degree of distortion, the error in wave heights owing to the effect not modeled may not be large.

Moveable-Bed River Models

Note that portions of the information in this section have been adapted from Pokrefke et al. (2012).

Physical movable-bed (or mobile-bed) models for the design and layout of river training structures normally reproduce several kilometers of the river under study. The usual procedure in such a model is to locate the most critical portion of the reach near the center (upstream to downstream) in the model. For example, in controlled waterways, such a configuration would locate the lock and dam under study in the center of the model.

Some innovative or uncommon river training structure types, such as vane dikes and bendway weirs, were first investigated in physical models before being installed in various river systems. Those investigations were important in developing design parameters for such structures to ensure that the structures would perform satisfactorily and maintain safe and efficient navigation conditions when installed in the prototype.

The size or scale of physical movable-bed models varies somewhat depending on the river under study, the type of movable-bed material used, and space or model discharge available. Many riverine movable-bed models have been constructed at scales varying from 1:60 (model to prototype) to 1:500 or greater. Most movable-bed river models were distorted in that the vertical scale was larger—1:60 or 1:80—depending on the movable-bed material used. Fine-grained sand or crushed coal were used on many movable-bed models; although more exotic materials such as walnut shells and chemically treated sawdust have been employed. Model sand usually has a specific gravity of 2.65 with a median grain size of about 0.25 mm, and model coal has a specific gravity of about 1.3 with a median grain size of 2 to 4 mm. A model constructed using sand must be more vertically distorted because the model bed material used is essentially the same, as the prototype and velocities to move the model bed material had to be similar to prototype velocities. With a sand bed model, the horizontal scale may often be 1:300 and the vertical scale 1:60. A model constructed using coal does not require as much distortion of the velocity to move the bed material; therefore, such models have often been constructed at undistorted scales of 1:100 or 1:120 for a reach on a large river such as the Mississippi or Ohio River. On smaller rivers, such as the Arkansas or Red River, the horizontal scale may have been 1:120 and the vertical scale 1:80.

It should be noted that even if a physical movable-bed model is constructed to an undistorted scale, inherent distortions still exist in such models. Typically, the bed slope from upstream to downstream is increased to obtain bed material movement. In addition, the velocity scale is typically increased over the theoretical Froude values to obtain bed material movement, particularly on medium to low flows. These distortions or departures from the theoretical values mean that velocities taken on the model cannot be meaningfully associated with prototype velocities. Also, these distortions tend to elevate the stages in the model; therefore, a physical movable-bed model cannot be used to accurately evaluate the impacts of river training structures on river stages. At best, stage readings can only indicate trends (increases or decreases) in stages as a result of the addition of river training structures and the resulting channel bed configuration.

Micromodels form a separate class of river sedimentation physical moveable bed models. Micromodels employ Froude number inequality between prototype and model and large linear scale distortions to greatly reduce space and cost requirements (Clancey et al. 2013). Maynord (2006) concluded that micromodels lacked quantitative predictive capabilities and recommended that they be used only for demonstration, education, and communication purposes.

The ASCE *Hydraulic Modeling: Concepts and Practice*, MOP 97 (Ettema 2000) provides an excellent resource on river modeling, as does Franco (1978) concerning details to be addressed during physical, movable-bed model studies.

Movable-Bed Coastal Models

A number of other parameters for defining similitude conditions are potentially important to the coastal movable-bed model. The similarity of grain—size distributions is usually ignored, but in situations where grain sorting plays a significant role, it should be considered. The ratio of grain density to water density affects sediment particle trajectories. Wave characteristics such as orbital velocity, orbit size near the bottom, and wave period can be combined to form several dimensionless groups including wave versions of F_* and R_*. Wave period compared with particle fall velocity can be an important consideration when waves contribute significantly to the transport process. These and other parameters are of potential importance, and various authors advocate different combinations to define similitude conditions. The choice of which parameters to use is dependent on what the modeler perceives as the dominant transport mechanism and practical constraints upon model design.

Equations (5-9) and (5-10) or other suitable noncohesive sediment transport formulae can be used to generate similitude conditions for modeling. The result is usually a model sediment that is larger and less dense than the prototype sediment (e.g., Letter and McAnally 1977), but strict adherence to the results is not always necessary for satisfactory modeling results. In many cases, it may not be possible to satisfy equational similitude with any practical model sediment. For example, model geometric scales dictated by cost and characteristics of the site to be modeled may produce a scaled sediment that is so light that embedded microbubbles cause particle floatation or transport at low flow shear stresses. Sediment transport equations can also be used to formulate sedimentation model timescales; however, sedimentation timescales are usually determined empirically.

Hudson et al. (1979) and Hughes (1993) provide a comprehensive guide to physical modeling of coastal processes, including design, construction, operation, and application to coastal deposition and erosion issues.

Reconciliation of scaling rules for tides, wind waves refraction, diffraction, and breaking, plus sediment behavior requires a thorough knowledge of sediment transport science and extensive experience.

Fixed-Bed Tracer Models

Fixed bed models, usually constructed of concrete or plastic, have been used with tracer materials to simulate sediment transport and deposition but not erosion. For example, Benson and Boland (1986) described model experiments using plastic granules in a fixed bed model of the Mississippi River Passes.

Gilsonite, a finely ground asphaltic material, has been used to simulate fine-grained sediment transport and deposition. The gilsonite technique does not reproduce flocculation of fine-grained sediments; therefore, it will not directly model sedimentation changes because of altered flocculation rates caused by geometry, structures, and other shear-producing factors. Those effects must be simulated by a sediment injection procedure that alters the supply available for deposition.

Physical Model Application Examples

Physical models also have a long history of evaluating sedimentation issues relative to navigation projects. Despite the lack of rigorously defined scaling relationships, such models have provided useful information to the engineer in the design, planning, and operation of navigation projects in both inland (e.g., Franco 1978) and coastal (e.g., Hudson et al. 1979) locations. Appendix B includes a case study for the physical model of the Dogtooth Bend reach, 20 mi above the Ohio and Mississippi Rivers confluence.

Simmons and Boland (1969) reported use of a moveable bed model of Galveston Harbor Entrance, built to scales of 1:100 vertical and 1:500 horizontal, to study channel sedimentation caused by both tidal currents, waves, and wave-driven currents under various channel realignment plans. Letter and McAnally (1977) compared those model predictions and observed prototype behavior for a 6-year period following completion of channel realignment. They concluded that qualitative prediction that undermining of the north jetty would be halted was correct and that the predicted relative increase in total maintenance dredging was reasonably close to that experienced in the prototype, although the patterns differed.

Herrmann and Tallant (1972) used gilsonite tracer to simulate fine-grained sediment transport and deposition in tidal Brunswick Harbor, Georgia. They tested two closure dams designed to reduce harbor channel deposition. Later, Letter and McAnally (1981) examined Brunswick Harbor prototype hydrographic surveys for approximately 6.5 years prior to construction of the first closure dam, 5.5 years between the first and second closure dams, and 1 year after the completion of the second closure dam. They found that the model correctly predicted the reduction and redistribution of shoaling with the first closure dam. However, with the second closure dam, the model predicted a large reduction in shoaling volumes, whereas the prototype experienced a slight increase in shoaling. They concluded that the prediction for the second closure dam was in error because of the difference in basic criteria for transport and deposition in the model and prototype (noncohesive versus cohesive, respectively) and to the drastic change in hydraulic conditions with the second closure dam relative to the shoaling validation period.

A movable-bed physical model of the Mississippi River Delta was constructed at Louisiana State University with 1:12,000 horizontal and 1:500 vertical scales to study large-scale diversions from the river. The model encompasses about 95 mi of the river from River mile 77.5 to the Gulf of Mexico at the end of Southwest Pass. From Froudian similitude the hydraulics timescale is 1/537. The sedimentation timescale, based on the times required to fill a known volume with sediment, is 1/17,857 (Willson et al. 2007). The model sediment consists of polystyrene pellets with a specific gravity of 1.05 and median grain size of 0.40 to 0.45 mm, yielding model to protype ratios of $d_{50,r} = 3.3$ and $\rho_{s,r} = 0.40$ (Cagle 2021).

NUMERICAL MODELS

Process-based numerical methods for estimating navigation channel sedimentation rates solve differential and algebraic equations representing the physical forcings and transport processes relevant to sedimentation. Such models include, at a minimum, submodels for hydrodynamics and sediment transport. Additional submodels which may be needed include those for wind waves, vessel traffic, and dredging.

Although the development of a numerical model is more data-intensive and requires more time and labor than the desktop approaches discussed previously, there are several key advantages to the use of a numerical modeling approach for estimating sedimentation rates. These advantages include the ability to resolve spatial and temporal gradients and the ability to solve for complex geometries and physical settings. In addition, numerical models can represent nonlinearities such as the impact on sediment transport from potential changes in hydrodynamics caused by thalweg meandering, channel deepening (e.g., a transition from ebb-dominant tidal currents to flood-dominance which can promote deposition), changes in the tidal prism and tidal currents caused by sea-level rise (owing to additional areas subject to tidal inundation), morphodynamics (the interplay between changing bathymetry owing to

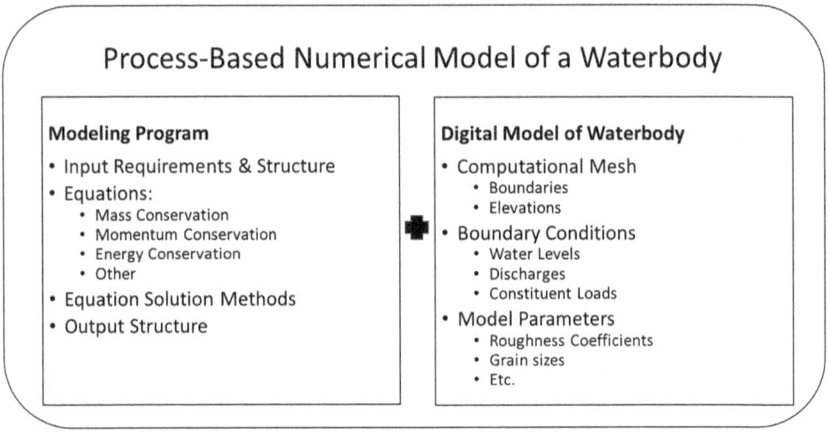

Figure 5-5. Components of a numerical waterway model.

sedimentation and hydrodynamics) which can affect the long-term temporal trend in sedimentation rate, and so on. In addition, numerical models also enable the development and detailed design of engineering solutions for managing sedimentation problems and provide information for economic and environmental analyses.

Numerical sedimentation models of waterways consist of two principal parts—a modeling program that solves the appropriate equations and a digital representation of the waterway to be modeled, as shown in Figure 5-5.

Numerical Model Programs

There are many numerical modeling programs; however, many are research tools, hardwired for specific conditions or constantly modified to test new theories and data. Others are proprietary and unavailable for use by anyone other than the developer. Production-level models are generalized for multiple site applications, tested to ensure mathematical sufficiency, peer-reviewed, and available for use by modelers other than the developers.

Numerical modeling programs solve equations for water and sediment mass and other constituent (e.g., temperature and salinity) conservation, momentum or energy conservation, and additional theoretical and empirical equations needed to perform calculations for transport, deposition, and erosion of sediment in waterways. The equations appear in many textbooks and program-specific user manuals. Garcia (2008) presents a comprehensive list of sediment transport equations and numerical model programs. The applicable equations are solved by iteration and approximate methods, including replacement of calculus differentials by finite differences or algebraic expressions and replacement of integrals by summations.

The programs solve the equations of motion and transport using finite difference, finite element, finite volume, and other methods on meshes (or grids) that represent the waterway. (The term *grid* suggests a rectangular system of points and *mesh* an irregular system. The two terms are considered interchangeable in this manual.) There has been much debate over which numerical methods are superior; however, any thoroughly tested and peer-reviewed model is capable of providing useful information to the skilled user with applicable expert sedimentation knowledge.

A program's solution method dictates the format of the digital waterway model, which is described next.

Digital Models of Waterways

The digital representation of a waterway includes inputs such as

- Mesh that defines the physical boundaries of the waterway and initial elevations of the bed, banks, structures, and overbanks;
- Some combination of water, sediment, and sometimes constituent inflows, outflows, and water levels at the water boundaries;
- Bed and inflowing sediment characteristics;
- Other driving forces information pertinent to the environment, such as wind, waves, precipitation, and air temperature; and
- Parameters required by the computer program equations such as boundary roughness, turbulence closure coefficients, sediment characteristics, and so on.

Model Dimensions

The number of spatial dimensions—zero-dimensional (0D), one-dimensional (1D), 2D, or 3D—constitutes a primary classification for numerical model program capabilities and applications. More spatial dimensions result in more equations to solve and that typically requires more computer time and storage space. Most sedimentation models perform time-varying computations. Bruner et al. (2020) describe the tradeoffs among 1D, 2D, and 3D hydrodynamic models and the relative hydrodynamic capabilities of each.

Modeling programs use several terms for discretization of the computational domain, including grid, mesh, or cell for geometric objects representing area and volume, and computation locations (points, intersections, or nodes) within the model domain. Cross sections serve as the mesh for 1D models. Terms tend to be associated with specific solution methods.

Models designated as 0D are box models with an inflow, outflow, and a calculated deposition/erosion process within the box. The box either accumulates or loses sediment mass. They are typically used in sediment budget tools and in lake/reservoir modeling.

Production-level 1D programs integrate hydrodynamic and transport processes over the vertical and lateral directions, that is, cross sections of a waterway, and calculate variations along the direction of flow. Bed exchange occurs over a cross section or a channel reach.

Vertical 2D (laterally averaged) programs (2DV) integrate the equations of motion and transport over width such that hydrodynamic and transport processes vary in two directions—longitudinally and vertically—over the modeled domain. Bed exchange occurs at the bottom water column cell as an average over the cell or as depth changes at each corner. These models were once common for stratified flow environments (e.g., Smith et al. 1987) but have fallen out of use in sedimentation studies with the advent of efficient 3D models. They are still employed in water quality modeling studies.

Horizontal 2D (vertically averaged) programs (2DH) integrate the equations of motion and transport over waterway depth such that hydrodynamic and transport processes vary in two directions—longitudinally and laterally. Bed exchange occurs at every cell in the computational mesh and is applied to the entire water column.

Production 3D programs solve a form of the 3D equations of motion and transport over waterway width, depth, and length. Most production programs employ the hydrostatic form of the equations of momentum—the vertical momentum equation is simplified by neglecting acceleration terms—to reduce computational burden. That approximation is appropriate in most navigation channel settings; however, it can produce errors in near-field calculations next to structures such as bridge piers, spillways, and other locations where vertical flow accelerations are significant. In such locations a 3D model without the hydrostatic approximation (3DN) may be required. Bed exchange occurs at the lowest cell in the computational mesh and is applied to the area encompassed by that cell or to each of the corner points.

Model Meshes

Digital model meshes, or grids, are networks of points and areas or volumes used to provide a geometric framework for which variables can be calculated and stored. They can be classified as either structured or unstructured. In a structured grid, the cells are regularly ordered such that they can be numbered in a sequential manner. They take the shapes of parallelograms (1D and 2D) or cuboids (3D) and the sides may be curved or straight. Figure 5-6 illustrates a 2D structured grid with curved sides. Structured grids are most often used in open ocean models where irregular boundaries do not constitute a significant part of the environment.

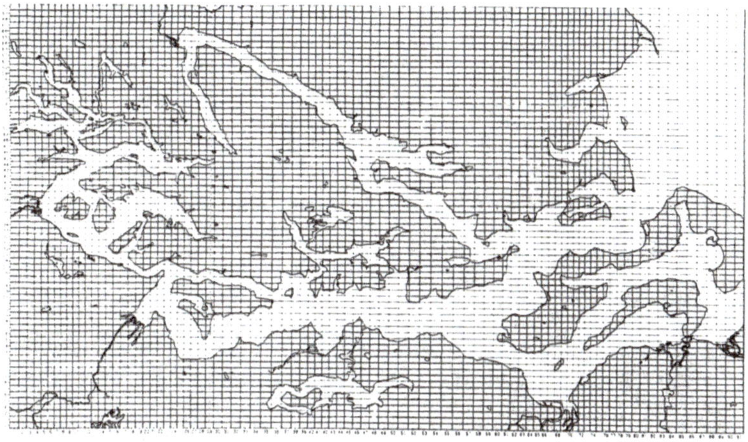

Figure 5-6. Example of a structured computation grid.
Source: Schmalz (1986).

Unstructured meshes consist of a variety of shapes based on triangles and rectangles and those shapes extended into three dimensions. The absence of inherent structure requires an indexing scheme that shows how each cell meets its neighbors. Unstructured meshes are the most common type for channel model studies. Figure 5-7 shows an unstructured mesh.

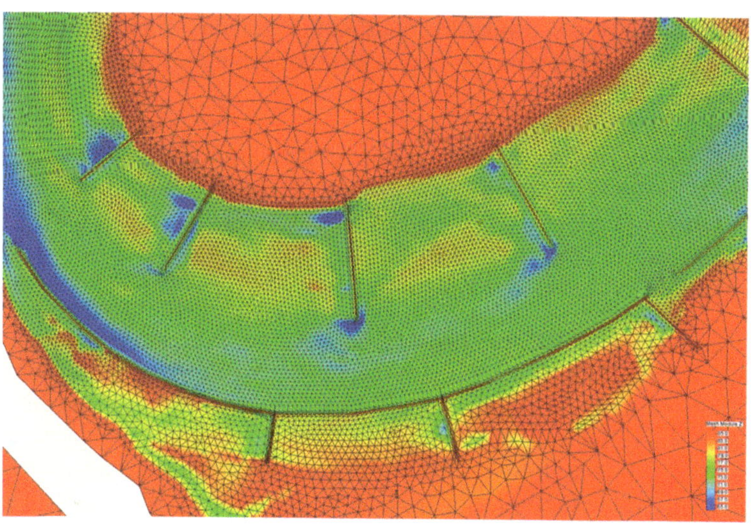

Figure 5-7. Example of an unstructured mesh of a river bend with Dike Field. Colors indicate bed elevations.
Source: Courtesy of Dynamic Solutions, LLC. Used with permission.

Topographic data are required if normally dry land may be inundated during model simulations. Waterway bottom and topography elevations are usually defined in a digital elevation model (DEM) as described in the previous section *Hydrographic Surveys*. They are used to define the land-water interfaces and bed elevations in a digital model mesh.

Some 1D models, such as HEC-RAS and HEC-6T, use meshes consisting of cross sections. DEMs can also be used to define these cross sections.

Model Programs Capabilities

Available numerical model programs offer a variety of capabilities that can be matched to the needs of a modeling project. Model features that may be useful include

- Graphical user interface,
- Wave generation and propagation,
- Nonhydrostatic equations for flows with strong vertical acceleration,
- Adaptive meshing and time stepping to refine results when needed,
- Dredging and dredged material placement,
- Vessel wake and propwash effects on sediment erosion,
- Morphological time step to reduce simulation times,
- Bendway correction term for 2DH models,
- Multiple grain size transport calculations,
- Tracer modeling,
- Nonequilibrium transport functions for noncohesive sediment,
- Flocculation processes functions for cohesive sediment,
- Bed consolidation,
- Bed armoring,
- Choice of constitutive equations and inherent parameters to model the sediment transport processes in the water column and the sediment bed, and
- Integrated modeling framework that includes hydrodynamics, waves, and sediment transport.

Because of the scale issues described in Chapter 2 section "Sedimentation Scales," computations of sediment transport can become burdensome for very long simulation periods. For example, a 1D model can easily simulate a decade of sedimentation, but a 3D model may require excessive computer time for a decade-long simulation. Such long-term sedimentation simulations are sometimes addressed by using algebraic simulations of morphological evolution over extended timescales in a more computationally efficient fashion (e.g., Donnell et al. 1991, Lesser et al. 2004). See "Time Step" section in Chapter 6 for further discussion of that capability.

Typical program choices based on capability include

- 1D models for
  - Constant water surface elevation on a given cross section;
  - Uniform velocity on a given cross section;
  - Unidirectional flow;
  - Negligible exchange of momentum between main channel and floodplain;
  - Sediment transport, deposition and erosion by river reach; and
  - Long (order of 1,000 km) reaches and simulation durations (decades).
- 2DH models for
  - Variable water level and velocity in longitudinal and lateral directions,
  - Nonuniform flow patterns,
  - Complex channel and floodplain interaction,
  - Different flow directions in low versus peak flows,
  - Insignificant vertical flow acceleration,
  - Insignificant vertical stratification,
  - Depth-average velocity,
  - Small secondary current effects,
  - Sediment transport plus bed erosion and deposition variation in longitudinal and lateral directions, and
  - Large modeled areas (order of 100 km) and moderate simulation durations (years).
- 3D hydrostatic models for
  - Stratified flow,
  - Secondary flows,
  - Complex flow regime with eddies,
  - Significant vertical gradients in suspended sediment concentration, and
  - Moderate modeled areas (order of 10 km) and simulation duration (months to years).
- 3D nonhydrostatic models for
  - Areas with large vertical accelerations,
  - Complex flow regimes, and
  - Small modeled areas (order of meters) and short simulation duration (hours).

A number of sedimentation numerical model programs with some or all these capabilities are available for use on navigation channel sedimentation problems. A partial listing of programs are shown in Table 5-3, classified by the number of spatial dimensions. The names of the models are usually acronyms, which are explained in Appendix D.

Selection of a modeling program is discussed further in Chapter 6, "Select Modeling Program."

Table 5-3. Example Numerical Model Programs for Navigation Channel Sedimentation.

Model	Primary use	References
1D models		
GenCade	Coastline Region	USACE (2021)
HEC-6T	Inland	MBH2O (2021)
HEC-RAS 1D	Inland	USACE (2020a)
2DH models		
AdH 2D/SEDLIB	Inland and Coastal	Brown (2012a, b)
CCHE2D	Inland and Coastal	Duan et al. (2001)
CMS	Coastal	Ding et al. (2007)
FUNWAVE-TVD	Coastal and channels	Demirbilek and Rosati (2011) Malej et al. (2019)
HEC-RAS 2D	Inland	USACE (2020b)
SRH-2D	Inland	Lai (2020)
XBeach	Coastal	Zimmermann et al. (2012)
3D hydrostatic models		
AdH 3D/SEDLIB	Inland and Coastal	Savant et al. (2010) McAlpin et al. (2020)
CCHE3D-SED	Inland and Coastal	Jia et al. (2002)
Delft3D	Inland and Coastal	Lemke et al. (2020)
ECOM	Inland and Coastal	Blumberg and Mellor (1987) Mathew and Winterwerp (2022)
EFDC	Inland and Coastal	EPA (2021)
Telemac	Inland and Coastal	Villaret et al. (2013)
FVCOM	Inland and Coastal	Chen et al. (2013) Wu et al. (2011)
MIKE	Inland and Coastal	Moharir et al. (2014)
3D nonhydrostatic models		
FLUENT	Inland and Coastal	Kaidi et al. (2021)
OpenFOAM	Inland and Coastal	Chen et al. (2014) Ouda and Toorman (2019)

Active numerical models tend to be under constant development to include improved physical knowledge and numerical algorithms, so potential users must check the current capabilities of any programs considered. Many of the models listed are paired with GUI to ease creating computational meshes and displaying and interpreting model results. Those interfaces are described in the following section.

Graphical User Interfaces

Some aspects of model use, most specifically generating, modifying, and viewing meshes and visualizing model results, are best performed by means of GUI. These interfaces allow these common modeling tasks to be performed rapidly and almost automatically. Most GUI are designed to work with one or a few model types; however, they also may allow importing and exporting files compatible with other models. Because they change so rapidly, this manual does not attempt to describe them in detail. Instead, some examples of their graphical displays are shown in Figures 5-8 and 5-9 to illustrate typical capabilities.

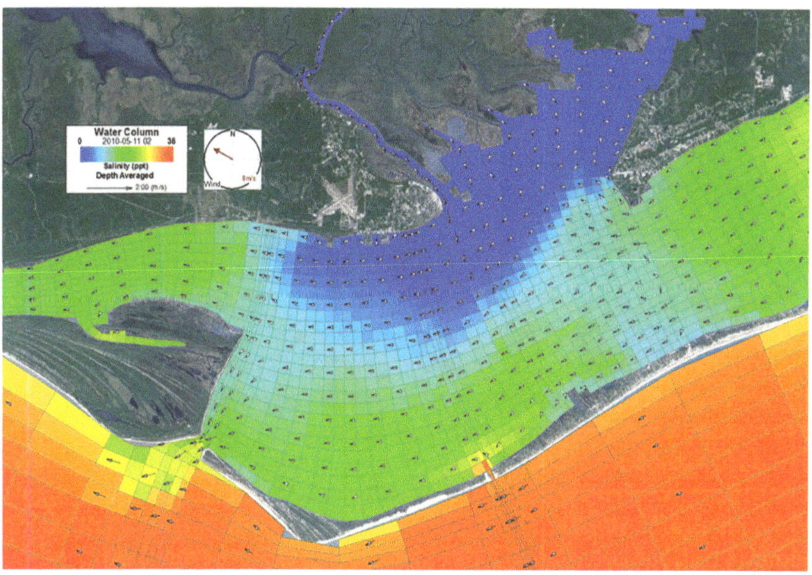

Figure 5-8. GUI Sample: EFDC_Explorer snapshot of Apalachicola Bay, Florida, salinity and depth-averaged current speed.
Source: Dynamic Solutions, LLC. Used with permission.

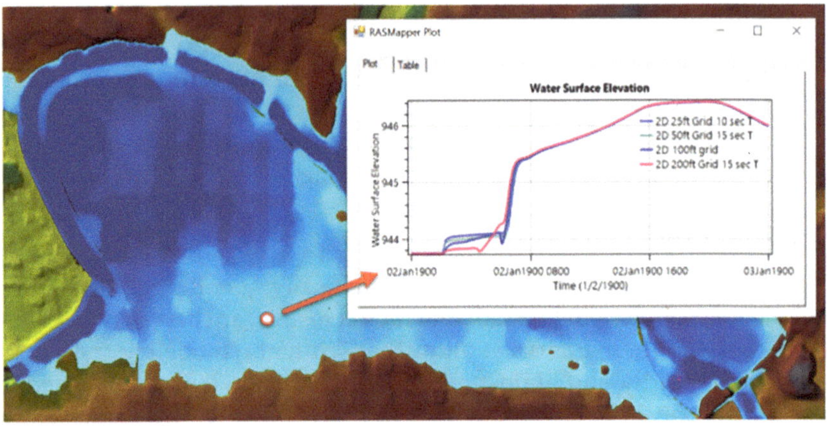

Figure 5-9. GUI Sample: HEC-RAS Mapper example display.
Source: HEC (2022).

UNCERTAINTY IN PREDICTIONS

All quantitative predictions of natural systems, including sedimentation, have inherent uncertainty resulting from errors in

- Inability to know future conditions (aleatory uncertainty).
- Our understanding and measurement of the physical and biological processes (epistemic uncertainty).
- Approximations and simplifications of the prediction method (structural uncertainty).

Epistemic and structural uncertainty applies to all models. The modeling programs listed in Table 5-3 usually employ the most recent advances in scientific understanding and numerical methods but uncertainty remains. Engineering decisions must be made despite these uncertainties, using the best available information with a full appreciation of the uncertainty bounds on that information. Therefore, modelers are responsible for quantifying uncertainty in their results and reporting them alongside model results (NAP 2012, EPA 2009). The following applies to all prediction methods but, for convenience, is expressed in terms of modeling.

The US Supreme Court recognized the importance of quantified uncertainty by means of the Daubert standard for admission of expert testimony into legal proceedings. The standard suggests that among the factors to be considered is the "... known or potential rate of error ..." in the methods applied (SCOTUS 1993). The science and engineering community has embraced these factors to varying degrees, with standards recognized by several journals and explained in detail by Wang et al. (2008). For numerical models of waterways, they include ensuring that the model solves the specified equations correctly—commonly called verification—and documented by the model developer in a peer-reviewed publication and ensuring that the model is applicable to the problem—commonly called validation—and documented for application of the model.

Sánchez et al. (2015) provide an overview of modeling uncertainty and definitions of some key terms.

Aleatory Uncertainty

Aleatory uncertainty—natural variability in future boundary conditions, including water level, meteorology, discharge, and sediment yield—presents the challenge that those conditions cannot be predicted deterministically. Hydrologic variables were traditionally treated as statistically stationary—individual values were random, but their probability distribution did not change. The long term mean and variance could be assumed constant. For shorter time periods when the mean was changing, such as known wet–dry climate cycles, time series could be made stationary by detrending or adopting a cyclostationary correction.

Hydrologic stationarity underpinned hydrologic and hydraulic engineering and thus sedimentation engineering for decades until large-scale human alterations to the landscape—urban, industrial, agricultural, and transportation development—changed land cover and rainfall-runoff-sediment yield and hydromodifications—dams, levees, channels, and jetties—changed waterway flows, slopes, waves, and sediment supply and character (e.g., Hovenga et al. 2016, Warrick 2015).

Beyond obvious human-constructed changes to the environment, a challenge to stationarity appears in the form of global warming and the climate change it produces (Arias et al. 2021). Milly et al. (2008) stated that "Stationarity is Dead" because of climate change and advocated a modified water management framework based on the recognition that climate will continue to change in unexpected, nonlinear, nonstationary ways. They recommended actions such as improved observation networks, enhanced climate models, and an effort to produce that framework; however, it does not yet exist.

Critics (e.g., Koutsoyiannis and Montanari 2015) of the "Stationarity is Dead" paper argued that stationarity can still be used for some components of natural processes but recognized that the basic theme is true—all water science and engineering faces increased uncertainty about future conditions, including those affecting sedimentation processes. Modelers must be aware of the increased uncertainty and they have a responsibility to communicate that awareness to anyone who uses their model results to make decisions.

Stationarity changes require sedimentation modelers to consider questions such as these when using observed data for future boundary conditions:

- What future basin land use/land cover changes will occur?
- What structural changes (coastal protection, dikes, dams, etc.) will be built?
- What effect will global warming have on hydrology and sea level?
- What effect will local and regional subsidence have on relative sea level and water levels?

If such changes are probable, they must be acknowledged and modeled as needed. Modeling these changes may require additional modeling efforts, such as watershed or regional models that account for the effects of future system changes.

USACE (2019) recognized the uncertainty associated with future rates of sea-level change. For planning studies and engineering designs over the project life, USACE guidance states that alternatives will be evaluated for a range of possible future rates of sea-level change, as represented by three different rates.

Epistemic and Structural Uncertainty

Uncertainty owing to limits of understanding and measurement of the physical and biological processes and approximations and simplifications of the solution method is caused by

- Hydrodynamic modeling errors are amplified (squared or cubed) in many sediment transport calculations.
- Sediment transport equations are mostly empirical, with sometimes difficult to define coefficients and large uncertainty bounds.
- Sediment data observations may be difficult and expensive to obtain. For these reasons they are often sparse and incomplete.
- All observed data contain errors and are subject to errors larger than nominal instrument error given by manufacturers.
- Sediment data for validation must be of the kind needed to satisfy the model's objectives. For example, validation to suspended sediment concentrations is useful but it does not show that the model is also validated to sediment deposition. Deposition depends on several independent factors, including particle size distribution, critical stresses, and bed shear stresses, not just concentration.

Model results must be interpreted in light of these uncertainties. Model validation and sensitivity testing are essential to proper interpretation, as discussed in Chapter 6.

Uncertainty Metrics from Observed Data

The ideal modeling approach employs multiple sets of good field observations to determine uncertainty bounds on the model results (NAP 2012, EPA 2009). There is extensive literature on estimating model uncertainty, sometimes called model skill assessment, using a variety of mathematical approaches. Statistical measures such as mean error [Equation (5-11)] and root mean square error (RMSE) [Equation (5-12)] are often used to quantify the differences between model result and observed data

$$\mu_e = \frac{\sum_{i=1}^{N}(x_o - x_m)_i}{N} \tag{5-11}$$

$$\sigma_e = \frac{\sum_{i=1}^{N}(x_o - x_m)_i^2}{N} \tag{5-12}$$

where N is the number of data comparison points, x is the variable of interest, and subscripts m and o indicate modeled and observed values, respectively. The RMSE measure is convenient for expressing model uncertainty bounds, because it is analogous to the statistical standard deviation about the mean, containing about 64% of the error within $\pm\sigma_e$ and 90% within $\pm 2\sigma_e$. Mean error and RMSE can be nondimensionalized by dividing them by the range or maximum value of x.

The Brier-Skill Score (BSS) for bed level as defined by Sutherland et al. (2004) and van Rijn et al. (2003) is given by

$$\text{BSS} = 1 - \left[\frac{\sum_{i=1}^{N}(Y_m - Y_o)_i^2}{\sum_{i=1}^{N}\Delta Y_o}\right] \tag{5-13}$$

where Y is the bed level, subscripts o and m indicate observed and model, respectively, and ΔY_o is the difference between beginning and ending bed levels. The BSS compares the mean square difference between the calculated and the measured bed levels with the mean square difference between the initial and the measured bed levels. A BSS of 1 indicates a perfect match whereas decreasing BSS indicates increasing divergence between the model and data. The BSS can be calculated over the resolution of the individual model grid cells resulting in a spatially comprehensive assessment; it may also be applied in a spatially segregated manner such as separately for areas inside and outside the navigation channel. The BSS has been used extensively by various authors to assess the performance of morphological models over various timescales (Ganju et al. 2009, van der Wegen and Jaffe 2014, Dam et al. 2016, Mathew and Winterwerp 2022). Van Rijn et al. (2003) provide guidance on the qualifying model performance on the basis of calculated BSS—values between 0.8 and 1 are categorized as "excellent," between 0.6 and 0.8 as "good," between 0.3 and 0.6 as "reasonable/fair," between 0 and 0.3 as "poor," and <0 as "bad."

The Murphy and Epstein (1989) Skill Score for any variable of interest, given as follows:

$$\mathrm{MESS} = 1 - \left[\frac{\sum_{i=1}^{N} \left(x_o - x_m \right)_i^2}{\sum_{i=1}^{N} \left(x_{0,i} - \bar{x}_0 \right)^2} \right] \tag{5-14}$$

where the overbar indicates the average value, offers a dimensionless metric varying from 0 (no agreement) to 1 (perfect agreement) with similar descriptions for intervening values as the BSS.

If statistical significance tests are used in model validation, care must be taken to express them properly. For example, it is never correct to say that a fact has been established as "statistically significant." Significance must be expressed as an exceedance probability, as in, "statistically significant to a 95% probability." Statistical *p*-values should not be used at all, as they are usually misused and almost universally misunderstood. Some scientific journals have banned their use (Nuzzo 2015).

Equations (5-11) through (5-14) provide estimates of model deviation from observed (field) values. As described in the *Field Observations* section of this chapter, observed data also contain error. To obtain the total RMS uncertainty, σ_T, these two sources of uncertainty must be combined through quadrature

$$\sigma_T = \sqrt{\sigma_F^2 + \sigma_e^2} \tag{5-15}$$

where σ_F is the RMS error in the field observations, and σ_e is the model RMSE from Equation (5-12).

Thorough validation-derived uncertainty, such as RMSE, can be used directly for base-to-plan comparisons which do not produce large changes in system responses. (In base-to-plan comparisons, the model simulation conditions are exactly the same between the base model and the plan model except for introduction of a management plan, such as freshwater diversion. Base-to-plan comparisons are considered ideal use of a model in that changes can be expressed in terms of relative trends and rates, not absolute values. The approach is described further in Chapter 6 section, *Testing*.) Our rule of thumb is that base-to-plan bed elevation changes of more than 30% require additional uncertainty allowance.

For other situations, including limited or no observed validation data, large changes, or different boundary conditions (e.g., future-casts) sensitivity simulation results—varying input parameters systematically and noting differences in model results—can be used as described in the following section.

Uncertainty with Limited Observed Data

In cases where observed data are not available to calculate uncertainty, other methods must be applied. The *Sensitivity Testing* section of Chapter 6 provides one approach for reducing and quantifying known sources of uncertainty in boundary location and mesh resolution.

Figure 5-10 illustrates a systems view of how uncertainty propagates through models. Each model uses three types of input to make computations and produce outputs that provide inputs to the next model. Each of the inputs—observed data (O), prior model outputs (MI), and parameter values (P)—exhibits uncertainty, each characterized by a probability distribution function (PDF), illustrated by the schematic curves

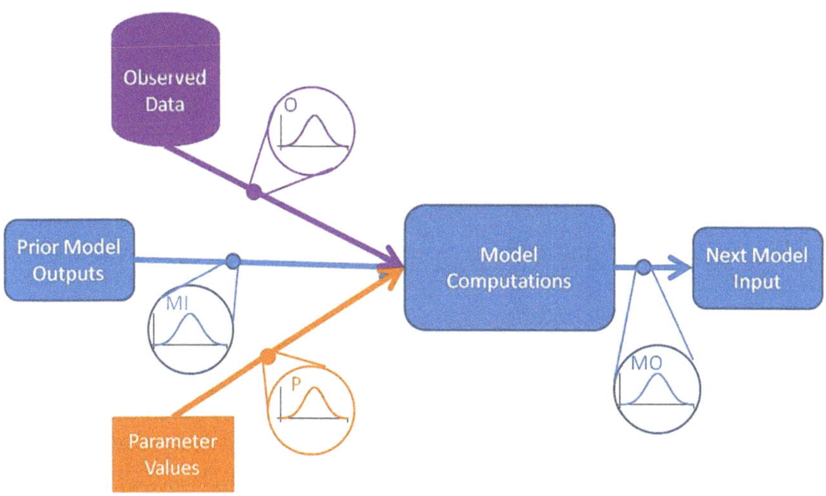

Figure 5-10. Systems view of uncertainty propagating as probability distribution from information sources through models.

in the bubbles. Those uncertainty distributions are propagated through the model computations and produce outputs with a different PDF (MO in the figure) that flow into the next model. Systems analysis with joint probability functions can be used to capture and compute the resulting outputs' probability distributions. See Nachlas (2012) and Ochi (1990) for a detailed description of joint probability calculations.

The generalized likelihood uncertainty estimation (GLUE) method provides estimates of uncertainty arising from parameter specifications. Widely used in water quality modeling, it has seen more limited use in sedimentation modeling, with the notable exception of Ruark et al. (2011), who applied a global sensitivity analysis, followed by a GLUE application and Latin hypercube sampling to a 1D noncohesive sediment transport model simulation of a flume experiment. They made 5,000-simulation tests, reduced to under 1,000 simulations in subsequent efforts, to examine the effect of eight input variables on model results.

The Monte Carlo approach randomly varies inputs to many sensitivity runs and produces a range of potential uncertainty as a PDF of model outputs. For steady-state results the process is relatively straightforward but random variable time series probabilities are best converted to stochastic probabilities (e.g., Savant 2008). Monte Carlo methods can be impractical for high resolution hydrodynamic and sedimentation process models because of computational limits. Point estimate methods offer a simpler, less computationally intensive way to estimate uncertainty than the Monte Carlo method. Pioneered by Rosenbluth and developed by Harr (1989), point estimate methods use three values—medium, high, and low values, usually taken to be mean and ±one or two standard deviations—for each input variable and then construct pseudo-PDFs from the model outputs by joint probability calculations. (e.g., Christian and Baecher 1999, Diaz-Ramirez 2007)

Reducing Uncertainty

Epistemic and structural uncertainty can be reduced with improved modeling, including higher dimensionality solutions, plus more and better observed data. The question becomes one of balancing time (cost) and uncertainty by clients and modeling managers.

Base-to-plan model testing, discussed previously and in Chapter 6, reduces uncertainty by comparing one model result to another and expressing the results as a relative change, such as percent decrease in reach deposition. Because both model simulations contain the same boundary condition uncertainties and structural error uncertainties, those effects are diminished somewhat when base-to-plan results are used. A commonly used rule of thumb is a 50% reduction in calculated epistemic and structural uncertainty. To a limited extent, base-to-plan results also mitigate aleatory uncertainty, because effects are presented as relative change, not absolute change. The degree of mitigation depends on the range of environmental conditions to which the model has been validated and tested. For example, a model tested with only significant wave heights (average of highest one-third) might improperly reproduce entrance channel sedimentation in both base and plan model tests.

REFERENCES

Abraham, D., J. Hendrickson, K. Jones, A. Jackson, et al. 2019. *Bedload transport measurements on the Chippewa River using the ISSDOTv2 method.* ERDC/TN RSM-20-3. Vicksburg, MS: USACE, Research and Development Center.

APHA (American Public Health Association). 1995. *Total suspended solids dried at 103°–105°C.* Method 2540 D. Washington, DC: APHA, American Water Works Association, and Water Pollution Control Federation.

Arias, P. A., N. Bellouin, E. Coppola, R. G. Jones, et al. 2021. "Technical summary." In *Climate change 2021: The physical science basis. Contribution of working group I to the sixth assessment report of the intergovernmental panel on climate change,* edited by V. Masson-Delmotte, P. Zhai, A. Pirani, S. L. Connors, et al., 33–144. Cambridge, UK: Cambridge University Press.

ASTM International. 2000. *Standard test method for determining sediment concentration in water samples.* ASTM D3977-97. West Conshohocken, PA: ASTM.

Benson, H. A., and R. A. Boland. 1986. *Mississippi River passes physical model study: Shoaling and hydraulic investigations in Southwest Pass.* Rep. 2. Tech. Rep. HL 86-1. Vicksburg, MS: USACE, Waterways Experiment Station.

Bhattacharya, B., R. K. Price, and D. P. Solomatine. 2007. "Machine learning approach to modeling sediment transport." *J. Hydraul. Eng.* 133 (4): 440–450.

Blumberg, A. F., and G. L. Mellor. 1987. "A description of a three-dimensional coastal model." In *Three-dimensional coastal ocean models,* edited by N. Heaps. Coastal and Estuarine Sciences Series Number 4. Washington, DC: American Geophysical Union.

Brown, G. L. 2012a. "A quasi-3D suspended sediment model using a set of correction factors applied to a depth averaged advection diffusion equation." In *Proc., IIHR 3rd Int. Shallow Flows Symp.* Iowa City, IA: University of Iowa.

Brown, G. L. 2012b. "Modification of the bed sediment equations of Spasojevic and Holly (1993) to account for variable porosity, variable grain specific gravity, and nonerodable boundaries." In *Proc., IIHR 3rd Int. Shallow Flows Symp.* Iowa City, IA: University of Iowa.

Brunner, G., G. Savant, and R. E. Heath. 2020. *Modeler application guidance for steady vs unsteady, and 1D vs. 2D vs 3D hydraulic modeling*, TD-41. Vicksburg, MS: USACE, Research and Development Center, Davis, CA: Hydrologic Engineering Center.

Bruun, P. 1988. "The Bruun rule of erosion by sea-level rise: A discussion on large-scale two-and three-dimensional usages." *J. Coastal Res.* 4 (4): 627–648.

Cagle, T. S. 2021. *Quantifying discharge on the lower Mississippi river physical model*. Baton Rouge, LA: Louisiana State University and Agricultural and Mechanical College.

Chen, C., R. C. Beardsley, G. Cowles, J. Qi, et al. 2013. *FVCOM user manual*. Cambridge, MA: Sea Grant College Program, Massachusetts Institute of Technology.

Chen, G., Q. Xiong, P. J. Morris, E. G. Paterson, et al. 2014. "OpenFOAM for computational fluid dynamics." *Notices Am. Meteorol. Soc.* 61 (4): 354–363.

Christian, J. T., and G. B. Baecher. 1999. "Point-estimate method as numerical quadrature." *J. Geotech. Geoenviron. Eng.* 125 (9): 779–786.

Clancey, K., S. Flash, R. Davinroy, J. Brown, et al. 2013. *Bumgard island HSR model river miles 36.0–25.0*. Tech. Rep. M65. St. Louis District, MO: US Army Corps of Engineers.

Corcoran, M. K., and J. R. Kelley. 2006. *Sediment-tracing technology: An overview*. ERDC TN-SWWRP-06-10. Vicksburg, MS: US Army Engineer Research and Development Center.

CTH (Committee on Tidal Hydraulics). 1995. *Reviews of problems in tidal waterways considered by the Committee on Tidal Hydraulics*. Washington, DC: US Army Corps of Engineers.

Dam, G., M. van der Wegen, R. J. Labeur, and D. Roelvink. 2016. "Modeling centuries of estuarine morphodynamics in the Western Scheldt estuary." *Geophys. Res. Lett.* 43 (8): 3839–3847.

Demirbilek, Z., and J. Rosati. 2011. *Verification and validation of the coastal modeling system*. Rep. 1: Summary Rep. ERDC/CHL TR-11-10. Vicksburg, MS: US Army Engineer Research and Development Center.

Diaz-Ramirez, J. N. 2007. "Assessment of uncertainty in flow model parameters, channel hydraulic properties, and rainfall data of a lumped watershed model." Ph.D. thesis, Mississippi State University, Rula School of Civil and Environmental Engineering.

Ding, Y., K.-C. Yeh, H.-K. Chen, and S. S. Y. Wang. 2007. "Simulations of morphodynamic changes due to waves and tides in an estuary using CCHE2D-coast model." In *World Environmental and Water Resources Congress*. Reston, VA: ASCE.

Donnell, B. P., J. V. Letter Jr., and A. M. Teeter. 1991. *The Atchafalaya River delta, Report 11: Two dimensional model*. Tech. Rep. HL-82-15. Vicksburg, MS: USACE, Waterways Experiment Station.

Dopsovic, R., L. Hardegree, and J. D. Rosati. 2002. *Sediment budget analysis system-A: SBAS-A for ArcView© application*. ERDC/CHL CHETN-XIV-7. Vicksburg, MS: USACE, Research and Development Center.

Downing, J. 2006. "Twenty-five years with OBS sensors: The good, the bad, and the ugly." *Cont. Shelf Res.* 26 (17–18): 2299–2318.

Duan, J. G., S. S. Y. Wang, and Y. Jia. 2001. "The applications of the enhanced CCHE2D model to study the alluvial channel migration processes." *J. Hydraul. Res.* 39 (5): 469–480.

Dunkin, L. M., L. A. Coe, and J. J. Ratcliff. 2018. *Corps shoaling analysis tool: Predicting channel shoaling*. ERDC/CHL TR-18-16. Vicksburg, MS: USACE, Research and Development Center.

EPA (US Environmental Protection Agency). 2009. *Guidance on the development, evaluation, and application of environmental models*. Washington, DC: EPA.

EPA. 2021. *Environmental fluid dynamics code (EFDC)*. Washington, DC: EPA.

Erickson, M. J., C. R. Barnes, M. R. Henderson, R. Romagnoli, et al. 2007. "Geomorphology-based interpretation of sedimentation rates from radiodating, Lower Passaic River, New Jersey, USA." *Integr. Environ. Assess. Manage.* 3 (2): 166–192.

Ettema, R. 2000. *Hydraulic modeling: Concepts and practice*. Reston, VA: ASCE.

Eysink, W. D., and H. Vermaas. 1983. *Computational methods to estimate sedimentation in dredged channels and harbour basins*. Publication No. 307. Delft, The Netherlands: Delft Hydraulics.

Feygels, V. I., J. Y. Park, J. Aitken, M. Kim, et al. 2012. "Coastal zone mapping and imaging LiDAR (CZMIL): First flights and system validation." In *Proc., SPIE 8532, Remote Sensing of the Ocean, Sea Ice, Coastal Waters, and Large Water Regions*. International Society for Optics and Photonics. Bellingham, WA.

Franco, J. J. 1978. *Guidelines for the design, adjustment, and operation of models for the study of river sedimentation problems*. Instruction Rep. H-78-1. Vicksburg, MS: USACE, Waterways Experiment Station.

Ganju, N. K., D. H. Schoellhamer, and B. E. Jaffe. 2009. "Hindcasting of decadal-timescale estuarine bathymetric change with a tidal-timescale model." *J. Geophys. Res* 114 (F4): F04019.

Garcia, M. 2007. *Sedimentation engineering: Processes, measurements, modeling, and practice*. Manuals and Reports on Engineering Practice No. 110. Reston, VA: ASCE.

Gray, J. R., G. D. Glysson, L. M. Turcios, and G. E. Schwarz. 2000. *Comparability of suspended-sediment concentration and total suspended solids data*. Water-Resources Investigations Rep. 00-4191. Reston, VA: US Geological Survey.

Gupta, D., B. B. Hazarika, M. Berlin, U. M. Sharma, et al. 2021. "Artificial intelligence for suspended sediment load prediction: A review." *Environ. Earth Sci.* 80 (9): 1–39.

Harr, M. E. 1989. "Probabilistic estimates for multivariate analysis." *Appl. Math. Modell.* 13 (5): 313–318.

Hayter, E., K. Gustavson, S. Ells, J. Gailani, et al. 2014. *Technical guidelines on performing a sediment erosion and deposition assessment (SEDA) at superfund sites.* Vicksburg, MS: USACE, Research and Development Center.

HEC (Hydraulic Engineering Center). 2022. *HEC-RAS mapper user's manual, version 6.2.* Davis, CA: US Geological Survey, Hydrologic Engineering Center.

Herrmann, F. A., and I. C. Tallant. 1972. *Plans for reduction of shoaling in Brunswick Harbor and Jekyll Creek, Georgia.* Tech. Rep. H72-5. Vicksburg, MS: USACE, Waterways Experiment Station.

Holmes, C. W. 1998. *Short-lived isotopic chronometers: A means of measuring decadal sedimentary dynamics.* USGS Fact Sheet 073-98. Washington, DC: US Geological Survey.

Hovenga, P. A., D. Wang, S. C. Medeiros, S. C. Hagen, et al. 2016. "The response of runoff and sediment loading in the Apalachicola River, Florida to climate and land use land cover change." *Earth's Future* 4 (5): 124–142.

Hudson, R. A., F. A. Herrmann Jr., R. A. Sager, R. W. Whalin, et al. 1979. *Coastal hydraulic models.* Special Rep. 5, Coastal Engineering Research Center. Vicksburg, MS: USACE, Waterways Experiment Station.

Hughes, S. A. 1993. Vol. 7 of *Physical models and laboratory techniques in coastal engineering.* Singapore: World Scientific.

Hughes, S. A. 2005. *Use of sediment trend analysis (STA®) for coastal projects.* Vicksburg, MS: USACE, Research and Development Center.

IHO (International Hydrographic Organization). 2020. *Standards for hydrographic surveys.* SR-44 Edition 6.0.0. Monaco: IHO.

Irish, J. L., J. K. McClung, and W. J. Lillycrop. 2016. *Airborne LiDAR bathymetry: The SHOALS system.* Washington, DC: US Army Corps of Engineers.

JALBTCX (Joint Airborne LiDAR Technical Center of Expertise). 2023. Accessed January 17, 2021. https://experience.arcgis.com/experience/f4c9c24ea6364a508ba7d9254060fdc1/page/Tools-%26-Data-Products/?views=Find-Data%2CData-%26-Web-Services%2CView-1.

Jia, Y., Y. Xu, and S. S. Y. Wang. 2002. "Numerical simulation of local scouring around a cylindrical pier." In *Proc., 1st Int. Conf. on Scour of Foundations.* Arlington, TX: Texas A&M University System, Texas Transportation Institute.

Kaidi, S., H. Smaoui, and P. Sergent. 2021. "Numerical investigation of the inland transport impact on the bed erosion and transport of suspended sediment: Propulsive system and confinement effect." *J. Mar. Sci. Eng.* 9 (7): 746.

Koutsoyiannis, D., and A. Montanari. 2015. "Negligent killing of scientific concepts: The stationarity case." *Hydrol. Sci. J.* 60 (7–8): 1174–1183.

Lai, Y. G. 2020. "A two-dimensional depth-averaged sediment transport mobile-bed model with polygonal meshes." *Water* 12 (4): 1032.

Lemke, L., M. S. Janssen, and J. K. Miller. 2020. "Mitigation of channel shoaling at a sheltered inlet subject to flood gate operations." *J. Mar. Sci. Eng.* 8: 865.

Lesser, G. R., J. V. Roelvink, J. T. M. van Kester, and G. S. Stelling. 2004. "Development and validation of a three-dimensional morphological model." *Coastal Eng.* 51 (8–9): 883–915.

Letter, J. V., and W. H. McAnally. 1977. *Physical hydraulic models: Assessment of predictive capabilities*. Research Rep. H-75-3. Vicksburg, MS: Galveston Entrance Movable-Bed Model, Waterways Experiment Station.

Letter, J. V., and W. H. McAnally. 1981. *Physical hydraulic models: Assessment of predictive capabilities*. Research Rep. H-75-3. Vicksburg, MS: Brunswick Harbor Shoaling Studies, Waterways Experiment Station.

Malej, M., F. Shi, and J. M. Smith. 2019. *Modeling ship-wake-induced sediment transport and morphological change—Sediment module in FUNWAVE-TVD*. ERDC/CHL CHETN-VII-20. Vicksburg, MS: USACE, Research and Development Center.

Mathew R., and J. C. Winterwerp. 2020. "Sediment dynamics and transport regimes in a narrow microtidal estuary." *Ocean Dyn.* 70 (2020): 435–462.

Mathew, R., and J. C. Winterwerp. 2022. "Morphodynamic modeling and morphological upscaling in a fine sediment system." *Adv. Water Resour.* 166 (August): 104224.

Maynord, S. T. 2006. "Evaluation of the micromodel: An extremely small-scale movable-bed model." *J. Hydraul. Eng.* 132 (4): 343–353.

MBH_2O (Mobile Boundary Hydraulics). 2021. https://www.mbh2o.com/hec-6t/.

McAlpin, T. O., J. V. Letter Jr., M. Bryant, A. G. Emiren, et al. 2020. *New York/New Jersey Harbor sedimentation study: Numerical modeling of hydrodynamics and sediment transport*. TR-20-15. Vicksburg, MS: USACE, Research and Development Center.

McAnally, W. H., C. Friedrichs, D. Hamilton, E. Hayter, et al. 2007. "Management of fluid mud in estuaries, bays, and lakes, part 2: Measurement, modeling, and management." *J. Hydraul. Eng.* 133 (1): 23–38.

Mehta, A. J. 2018. *An introduction to hydraulics of fine sediment transport*. 2nd ed. Singapore: World Scientific.

Midgley, T. L., G. A. Fox, and D. M. Heeren. 2012. "Evaluation of the bank stability and toe erosion model (BSTEM) for predicting lateral retreat on composite streambanks." *Geomorphology* 145–146: 107–114.

Mikkelsen, L., P. Mortensen, and T. Sorensen. 1980. "Sedimentation in dredged navigation channels." In *Proc., 17th Coastal Eng. Conf.*, 1719–1734. Reston, VA: ASCE.

Milly, P. C. D., J. Betancourt, M. Falkenmark, R. M. Hirsch, et al. 2008. "Stationarity is dead: Whither water management?" *Science* 319 (5863): 573–574.

Moharir, R. V., K. Khairnar, and W. N. Paunikar. 2014. "MIKE 3 as a modeling tool for flow characterization: A review of applications on water bodies." *Int. J. Adv. Stud. Comput. Sci. Eng.* 3 (3): 32–43.

Murphy, A. H., and E. S. Epstein. 1989. "Skill scores and correlation coefficients in model verification." *Mon. Weather Rev.* 117 (3): 572–582.

Nachlas, J. A. 2012. *Probability foundations for engineers*. Boca Raton, FL: CRC Press.

NAP (National Academies Press). 2012. *Assessing the reliability of complex models*. Washington, DC: National Academies of Science, Engineering and Medicine.

NOAA (National Oceanic and Atmospheric Administration). 2022. "Historical map and chart collection." Accessed September 1, 2023. https://historicalcharts.noaa.gov.

Nuzzo, R. 2015. "Scientists perturbed by loss of stat tools to sift research fudge from fact." *Sci. Am.* April 16. Accessed September 1, 2023. https://www.scientificamerican.com/article/scientists-perturbed-by-loss-of-stat-tools-to-sift-research-fudge-from-fact/.

Ochi, M. K. 1990. *Applied probability and stochastic processes. In engineering and physical sciences*. Wiley Series in Probability and Statistics. New York: Wiley.

Ouda, M., and E. A. Toorman. 2019. "Development of a new multiphase sediment transport model for free surface flows." *Int. J. Multiphase Flow* 117 (August): 81–102.

Parchure, T. M., W. H. McAnally Jr., and A. M. Teeter. 2001. "Desktop method for estimating vessel-induced sediment suspension." *J. Hydraul. Eng.* 127 (7): 577–587.

Pokrefke, T. J. 2012. *Inland navigation: channel training works*, MOP 124. Reston, VA: ASCE.

Prasadreddy, K. B. R., and M. Visweswararao. 2019. "Sedimentation in Sriramsagar project on Godavari-concerns and solution." *Int. J. Civ. Eng. Technol.* 10 (7): 22–28.

Pratt, T., and T. Parchure. 2003. "OBS calibration and field measurements." In *Proc., Federal Interagency Workshop on Turbidity and Other Sediment Surrogates*, edited by J. R. Gray and G. Douglas Glysson. US Geological Survey Circular 1250. Reno, NV: US Geological Survey.

Pugh, C. A. 2007. "Sediment transport scaling for physical models, Appendix C." In *Sedimentation engineering: Processes, measurements, modeling, and practice*, MOP 110, edited by M. Garcia. Reston, VA: ASCE.

Ramirez-Avila, J., E. Langendoen, W. McAnally, J. Martin, et al. 2010. "Assessment and estimation of streambank erosion rates in the Southeastern Plains ecoregion of Mississippi." In *Proc., Federal Interagency Sedimentation Conf.* Las Vegas, NV.

Rasmussen, P. P., J. R. Gray, G. Douglas Glysson, and A. C. Ziegler. 2011. "Guidelines and procedures for computing time-series suspended-sediment concentrations and loads from in-stream turbidity-sensor and streamflow data." Chap. 4, Book 3 in *Applications of hydraulics, section C, sediment and erosion techniques*. Techniques and Methods 3–C4. Reston, VA: US Geological Survey.

Robertson, D. E., and J. J. Fix. 1977. *Association of Hanford origin radionuclides with Columbia River sediment*. BNWL-2305. Richland, WA: Battelle Pacific Northwest Labs.

Ruark, M. D., J. D. Niemann, B. P. Greimann, and M. Arabi. 2011. "Method for assessing impacts of parameter uncertainty in sediment transport modeling applications." *J. Hydraul. Eng.* 137 (6): 623–636.

Sánchez, A., B. E. Skahill, W. Scharffenberg, and C. H. Smith. 2015. *Computer based model calibration and uncertainty analysis: Terms and concepts*. ERDC/CHL CHETN I-86. Vicksburg, MS: USACE, Research and Development Center.

Sarikaya, H. Z. 1977. "Numerical model for discrete settling." *J. Hydraul. Div.* 103 (8): 865–876.

Savant, G. 2008. "Prediction of estuarine morphological evolution." Ph.D. thesis, Mississippi State University, Rula School of Civil and Environmental Engineering.

Savant, G., R. C. Berger, T. O. McAlpin, and J. N. Tate. 2010. *Intelligent adaptive time-step control for modeling rapidly evolving hydrodynamic flows in adaptive hydraulics (ADH)*. Vicksburg, MS: USACE, Research and Development Center.

Schmalz, R. A. 1986. *A numerical model investigation of astronomic tidal circulation in Puget Sound*. Miscellaneous paper CERC-86-9. Vicksburg, MS: USACE, Waterways Experiment Station.

SCOTUS (Supreme Court of the United States). 1993. Daubert v. Merrell Dow Pharmaceuticals (92-102), 509 U.S. 579.

Scully, M. E., and C. T. Friedrichs. 2007. "Sediment pumping by tidal asymmetry in a partially mixed estuary." *J. Geophys. Res.* 112 (C7): C07028.

Sharp, J. A. 2007. "Sediment budget template applied to Aberdeen Pool." Master's *Thesis*, Mississippi State University, Dept. of Civil and Environmental Engineering.

Simmons, H. B., and R. A. Boland Jr. 1969. *Model study of Galveston Harbor entrance, Texas*. Hydraulic Model Investigation Tech. Rep. H-69-2. Vicksburg, MS: USACE, Waterways Experiment Station.

Smith, T. M., W. H. McAnally, and A. M. Teeter. 1987. *Corpus Christi Inner Harbor Shoaling Investigation*. Tech. Rep. HL 87-13. Vicksburg, MS: USACE, Waterways Experiment Station.

Sutherland, J., A. H. Peet, and R. L. Soulsby. 2004. "Evaluating the performance of morphological models." *Coastal Eng.* 51 (8): 917–939.
Sutherland, T. F., P. M. Lane, C. L. Amos, and J. Downing. 2000. "The calibration of optical backscatter sensors for suspended sediment of varying darkness levels." *Mar. Geol.* 162 (2–4): 587–597.
Trawle, M. J. 1981. "Effects of depth on dredging frequency: Report 2: Methods of estuarine shoaling analysis." 1–136. Vicksburg, MS: USAE Waterways Experiment Station.
Trawle, M. J., and J. B. Herbich. 1980. *Prediction of shoaling rates in offshore navigation channels*. CDS Rep. No. 232. College Station, TX: Center For Dredging Studies Dept. of Civil Engineering, Texas Engineering Experiment Station.
Tuell, G., K. Barbor, and J. Wozencraft. 2010 "Overview of the coastal zone mapping and imaging LiDAR (CZMIL): A new multisensor airborne mapping system for the US Army Corps of Engineers." In *Proc., SPIE 7695, Algorithms and Technologies for Multispectral, Hyperspectral, and Ultraspectral Imagery XVI*. International Society for Optics and Photonics. Bellingham, WA.
USACE (US Army Corps of Engineers). 1989. *Sedimentation investigations of rivers and reservoirs*. Engineer Manual 1110-2-4000. Washington, DC: USACE.
USACE. 2013. *Standards for hydrographic surveys*. EM 1110-2-1003. Washington, DC: USACE.
USACE. 2019. *Incorporating sea level change in civil works programs*. ER 1100-2-8162. Washington, DC: USACE.
USACE. 2020a. *HEC-RAS sediment transport user's manual*. Davis, CA: USACE, Hydrologic Engineering Center
USACE. 2020b. *HEC-RAS two-dimensional sediment transport user's manual*. Davis, CA: USACE, Hydrologic Engineering Center.
USACE. 2021. *Coastal inlets research program*. GenCade—v1.1 r8. Vicksburg, MS: USACE, Research and Development Center.
USACE. 2022. *Surveying and mapping*. Washington, DC: USACE.
USGS (US Geological Survey). 2007. *Guidelines for sampling bed material*. Technical Supplement 13A, Part 654, National Engineering Handbook. Reston, VA: USGS.
van der Wegen, M., and B. E. Jaffe. 2014. "Processes governing decadal-scale depositional narrowing of the major tidal channel in San Pablo Bay, California, USA." *J. Geophys. Res. Earth Surf.* 119 (5): 1136–1154.
van Rijn, L. C. 2013. *Principles of sedimentation and erosion engineering in rivers, estuaries and coastal seas*. Madison, WI: Aqua Publications.
van Rijn, L. C., D. J. Walstra, B. Grasmeijer, J. Sutherland, et al. 2003. "The predictability of cross-shore bed evolution of sandy beaches at the time scale of storms and seasons using process-based profile models." *Coastal Eng.* 47 (3): 295–327.

Villaret, C., J.-M. Hervouet, R. Kopmann, U. Merkel, et al. 2013. "Morphodynamic modeling using the Telemac finite-element system." *Comput. Geosci.* 53 (April): 105–113.

Wall, G. R., E. A. Nystrom, and S. Litten. 2006. *Use of an ADCP to compute suspended-sediment discharge in the tidal Hudson River*. SIR 2006-5055. New York: US Geological Survey.

Walling, D. E. 2013. "The evolution of sediment source fingerprinting investigations in fluvial systems." *J. Soils Sediments* 13 (December): 1658–1675.

Wang, S. Y., P. J. Roache, R. A. Schmalz, Y. Jia, et al. 2008. *Verification and validation of 3D free-surface flow models*. Reston, VA: ASCE.

Warrick, J. A. 2015. "Trend analyses with river sediment rating curves." *Hydrol. Processes* 29 (6): 936–949.

Willson, C. S., N. Dill, W. Barlett, S. Danchuk, et al. 2007. "Physical and numerical modeling of river and sediment diversions in the lower Mississippi River delta." In *Proc., Coastal Sediments '07*, 749–761. Reston, VA: ASCE.

Wu, L., C. Chen, P. Guo, J. Qi, et al. 2011. "A FVCOM-based unstructured grid wave, current, sediment transport model, I. Model description and validation." *J. Ocean Univ. China* 10 (1): 1–8.

Zimmermann, N., K. Trouw, L. Wang, M. Mathys, et al. 2012. "Longshore transport and sedimentation in a navigation channel at Blankenberge, Belgium." *Coastal Eng. Proc.* 1 (33) 1–16.

CHAPTER 6
BEST PRACTICES

Best practices for predicting channel sedimentation are expressed here in terms of models, with the emphasis on numerical models; however, these concepts can lead to best use of field data, desktop modeling, and physical modeling, as needed.

Best numerical modeling practices in general, including hydrodynamics, are widely available, notably in USACE (Brunner et al. 2020), the Environmental Protection Agency (EPA 2009), National Academy of Sciences, Engineering and Medicine (NAP 2012), and the Dutch Department of Civil Works (STOWA 1999) documents. Other excellent sources of general numerical modeling best practices include Wang et al. (2008) and Martin and McCutcheon (1999). In addition, the professional literature contains a wealth of information on successful numerical modeling.

Sedimentation problems are best solved through a methodical, step-by-step approach. The advent of easy-to-use numerical models has increased the temptation to immediately begin modeling without taking the necessary preparatory steps. The tendency to rapidly proceed with an intuitive solution may or may not produce good results, and too often requires re-doing work flawed by oversights. For example, the authors are aware of a lateral dike constructed along the west side of an offshore bar channel in the mistaken belief that longshore transport was from west to east. The sediment deposition problem worsened because net transport was actually from east to west. A marina sedimentation model study failed because the sediment source was misidentified. A river engineering solution required a redesign after ignored wash load became the primary problem. Such stories abound.

Here, we recommend a methodical approach to numerical modeling—the following steps are similar to those recommended by USACE (1989) for rivers and EPA (2009) for water-quality studies and are the product of our experiences.

PROBLEM DEFINITION

The sediment problem should be defined by the person(s) affected by it and those responsible for addressing it. Sometimes that is a single client but may involve several individuals. For example, the primary client in a port channel project will usually be the port engineer. Other involved stakeholders may be the port director and vessel pilots/operators. Larger projects may have a program and project or operations manager, staff engineer, channel users, and port and terminal officials.

A key aspect of the problem definition is the client's quality expectations, as expressed in precision of needed results and limits of time, cost, and other resources. For example, a preliminary study to reduce the number of plans under consideration allows approximate results that can be used to eliminate plans with minimal benefits or adverse effects. However, modeling a final plan for design and an Environmental Impact Statement requires more precise, reliable, and defensible results. This concept, expressed as "fitness for intended use" by Juran (1980) and "ability to satisfy stated or implied needs" by the International Organization for Standards (ISO 2015), underpins quality management plans for all projects, including modeling.

Problem definition should include the following.

Objectives Statement

A clear, unambiguous statement of what is to be achieved—the end result—is essential to focus the work and prevent misunderstanding. For many modeling studies, adding a list of specific questions is useful in sharpening that focus. Such questions might include

- Which of five proposed training dike plans will provide the least shoaling in the navigation channel?
- Is it economically and environmentally feasible to deepen the navigation channel?
- What model outputs are needed to provide input to the economic and environmental models?

Other issues affecting the objectives, such as expected quality, habitat conditions, presence of threatened/endangered species, or avoiding adverse currents, should be included along with the objectives. (See the "Fitness for Use" section that follows for a further discussion of quality objectives.)

Similarly, constraints on objectives or approach imposed by law or client needs should be stated if the client does not object. Some clients do not want modelers to say that time or costs are constrained.

Problem and Process Boundaries Delineation

Problem areas are usually defined by the client—a given reach of channel where sedimentation issues arise. However, reliable numerical model boundaries for the problem area must be chosen and tested to ensure that they are far enough removed from the problem area that boundary specifications do not interfere with model solutions in the area of interest. For some problems, the required spatial or temporal extent may require the use of multiple models. For example, the Atchafalaya Bay study described in Appendix B: Case Studies used regional models to simulate long-term Atchafalaya Basin and Gulf of Mexico conditions and then more highly resolved local models, including inset meshes to simulate delta building details.

Defining the process boundaries must also take into account the morphological processes described in Chapter 2 section "Sedimentation Scales." What morphologic and hydrologic trends are, or will be, affecting the region of the project site over its life span? Anticipated changes should be accounted for in locating boundaries and specifying boundary conditions, as discussed in a subsequent section.

Less obvious concerns are those areas outside the problem area but affected by changes within the problem area. For example, jetties designed to protect an inlet channel will influence sedimentation in the inlet channel and adjacent beaches but may also cause beach erosion far from the inlet. If those areas are not included in the evaluation, that aspect should be explicitly mentioned as one of the study's limitations.

Information Inventory

An inventory of available information, including a review of all prior studies and available databases, will establish which remedies may have been previously examined, what data are available, and where data gaps exist. Modelers should understand that useful information exists beyond that revealed by online searches. If data gaps are identified, a data collection plan should be designed to collect needed information and recommended in the proposed scope of work. The project objectives should be revisited and modified, depending on additional data collection needs. See the "Field Observations" section of Chapter 5, "Prediction Methods," for a list of typical data needs. Recommendations for field data to support modeling can be found in these environment-specific sources:

- Coastal—Hudson et al. (1979),
- Estuarine—USACE (1991), and
- Riverine—USACE (1989).

Conceptual Site Model

The conceptual site model is a qualitative description of "the most important behaviors of the system, object, or process relevant to the problem of interest" (EPA 2009). The conceptual site model can serve to summarize various lines of evidence and conceptualize the dominant processes relevant for sedimentation at the project site. It can also be supplemented by graphical representations of the fate and transport processes and thus support effective communication with the various stakeholders.

The conceptual site model can be developed using existing data from the site, historical literature or reports, or site knowledge. Examples of specific questions to be addressed as part of site model development include the following:

- What are the key fate and transport processes relevant for sediment transport, and consequently, sedimentation?
- How does sediment transport scale in response to the various hydrodynamic forcings?
- What are the spatial and temporal scales of relevance for the various transport processes?
- How do episodic events contribute to sedimentation?

Development of a conceptual site model as a precursor to the development of the numerical model is recommended in the literature (Winterwerp et al. 2021) and in guidance from various organizations (EPA 2009, STOWA 1999). Its development can also be a part of a feedback loop with the development of the numerical model, with information gained from applying the numerical model used to refine the conceptual site model in an iterative fashion.

A robust and well-characterized conceptual site model can help with the selection of the numerical modeling framework, define processes to be modeled, identify the scales of relevance for the model application which can determine the spatial and temporal resolution of the model, develop validation metrics, identify data gaps, support the design of field sampling programs, and so on. The conceptual site model can provide a qualitative framework for communicating model results and help with constraining and interpreting model performance, especially when applied to forecasting future scenarios, for example, in assessing the impact of any planned engineering works. Therefore, including the development of a conceptual site model as part of the project life cycle will ensure that relevant fate and transport processes have been appropriately incorporated into the numerical model and provide confidence in model projections for future scenarios.

APPROACH AND TOOLS SELECTION

Design Approach

Phased application of tools from simplest to most complex in a tiered approach provides several advantages. An example of a tiered approach to a channel deposition problem is to

Tier 1. Conduct a desktop analysis, such as the volume of cut method or a sediment budget as described in Chapter 5, "Prediction Methods."

Tier 2. Construct and test a simplified digital model (similar boundaries as the final but much lower resolution) of the problem area, using steady-state boundary conditions, comparing the results with those from the desktop calculations and checking for reasonableness.

Tier 3. Develop, validate, and test a full model, as described in the following.

The advantages of a tiered approach include a comparison of results from differing methods to identify major mistakes early, identification of geometry (boundary and depth) issues before embarking on time-consuming model validation and pretesting of boundary conditions for sufficiency at lower resolution/less computational cost. The benefits are faster completion of the work and fewer mistakes.

A tiered approach can reduce predictions' uncertainty by comparing results from multiple methods. For example, the Atchafalaya Bay delta predictions described in Appendix B used field, desktop, physical modeling and several one-dimensional (1D) and two-dimensional (2D) numerical models to compile multiple projections for 50-year delta development and combined them to provide an integrated set of predictions.

The tiered approach also fits well with the phases of a federal project. In the planning phase for federal navigation projects, analyses and modeling focuses on three aspects: Does the project/solution have engineering feasibility, economic justification, and environmental acceptability? During this phase, Tier 1 desktop analyses may be sufficient or Tier 2 modeling may be needed if the objective is to screen out alternatives, identify impacts, and find potential solutions. In the design phase, more complex modeling may be conducted to determine design parameters for development of plans and specifications and to finalize operational and maintenance requirements.

Identify Scale Issues

The sedimentation processes scale issues discussed in Chapter 2, "Channel Sedimentation Processes," intersect with model selection and boundaries identification to affect subsequent model application steps. For example, if sedimentation within a coastal or riverine dike field (scale on the order of yards) is a significant issue and planned improvements either affect, or will be affected by, possible changed conditions large distances (order of miles) away, the scale differences may indicate which and how many models are used. Successful modeling may require two or more digital models—nested or connected at boundaries—or two or more model programs, for example, a watershed model and a river model. Such questions must be addressed in concert with the modeling program selection discussed below. For example, a modeling program employing a structured grid may employ an excessive number of cells over a large, irregular area and require a nested grid approach to capture the proper scales in the problem area.

When channel sediment predictions are needed over the life of the project, which can have timescales of 50 to 100 years or more, successful modeling may require multiple mesh updates to reflect morphological changes over those timescales. This is especially true when multidimensional modeling is needed and computational capability is limited to months or a year or two because of factors such as processing time. Modeling can be performed in one- or two-year snapshots several times over the 50- to 100-year project life. This practice also allows for the incorporation of changing boundary conditions over time, such as sea-level rise. Techniques for accomplishing long-term simulations are described further in the "Time Steps" section and in Appendix B.

Assess Resources

Study design must include a candid assessment of resources available. Sedimentation numerical modeling requires computing resources sufficient to the computer model program selected and the size and resolution of the digital model used to describe the modeled area. Personnel knowledgeable in numerical modeling practice (separate from theory) can obviously accomplish modeling tasks more quickly and correctly, as can those equipped with modern graphical user interfaces (GUI) for mesh creation and modification and display of results.

Most important to the final product are personnel with a deep understanding of the physical processes involved—hydrodynamics and the appropriate (cohesive, noncohesive, or mixed) sediment transport. Such a thorough understanding of the processes of the modeled site is essential for the modeler to recognize physically implausible data and results. A knowledge of numerical modeling is necessary, but it is not sufficient.

Finally, schedule and funds available for the work must be compatible with the client's expected results, including quality. If resources are insufficient to fully meet client goals and good practice requirements as described here, less robust modeling (e.g., coarser resolution, less validation effort) can be performed with a greater range of uncertainty in results. See Chapter 5 section "Uncertainty in Predictions" for estimating uncertainty.

As an example of these trade-offs, the USACE Smart Planning (USACE 2015b) decision and product milestones influence modeling approach in the feasibility phase. The modeler/engineer must be able to address how uncertainties will be affected if a less robust modeling approach will substantially bias study outcomes, including what solutions are possible and what solutions are feasible, or whether stakeholder or client safety is affected by the choice of the model. There may be a need to take additional time to perform more rigorous modeling in the feasibility phase and not wait until the design phase.

Select Tools

Selecting the proper modeling program to predict channel sedimentation and evaluate remedies requires a thorough understanding of the site characteristics and hydrodynamic and sedimentation processes discussed in Chapter 2. In addition, it is necessary to identify the space and time scales involved, required level of accuracy, the modeling platform (i.e., computer capacity), time available, and budget. That knowledge guides the selection of tools. Chapter 5 section, "Model Program Capabilities" provides information on the various numerical modeling capabilities that should be considered.

It is not always possible, nor is it always necessary, to model all processes that affect sedimentation in a navigation channel. Some processes may have minimal influence on the solution or the modeling objective. For example, although the 1D models used by USACE to calculate shoaling in the Mississippi River's Southwest Pass (see Appendix B) lacked the ability to model the effect of salinity on shoaling, the team used their technical expertise to interpret the model results to estimate dredging needs. Those estimates were used in the economic analysis of alternatives for deepening the navigation channel. Those estimates would not have been sufficient for design purposes, but they met the planning study objectives.

Tool selection is addressed in more detail in Appendix C: Decision Tree Tool Selection.

All else being equal, a modeling program with which the modeler is familiar often becomes the best choice, particularly if time and cost are limited. Learning the peculiarities of a new modeling program consumes time that could be spent on testing.

Finally, the modeler's expertise is more important than the model choice. Interpretation, discussed in a following section, is key to providing sound advice to decision makers.

NUMERICAL MODELS APPLICATION

As described in the previous section, the recommended sequence of tool applications in a model investigation is three or more increasingly detailed tiers. This section is devoted to best practices in Tier 3—rigorous modeling of navigation channel sedimentation processes.

Successful modeling of sedimentation processes requires an excellent reproduction of flows – water levels, current speeds and directions, and/or short period waves, depending on the hydrodynamic environment. A hydrodynamic model validated to water levels alone is seldom sufficient for reliable sediment transport modeling.

For the purposes of this manual, a well-validated numerical model of hydrodynamics (verified modeling program and validated combination of modeling program and digital model) is assumed. Recommendations for modeling practice are limited here to sedimentation predictions in navigation channels with the assumption that hydrodynamic modeling of appropriate quality has been accomplished. See Wang et al. (2008) for a thorough discussion of hydrodynamic model validation.

Model Dimensionality and Mesh

Model dimensionality and resolution sufficient for the reproduction of significant sedimentation processes depends on the sedimentation environment, including the driving hydrodynamics, and questions to be addressed by the modeling. As discussed previously and described in Appendix C, the objectives, problems, and processes to be modeled and available resources will indicate the dimensionality (1D, 2D, 3D), resolution, extent, and number of digital model meshes. Typical application and resolution requirements are given in Table 6-1.

Resolution of the digital model mesh – size of the mesh cells – dictates the minimum dimensions of bathymetric features and flow/transport processes that can be accurately modeled. For example, a 100 m horizontal mesh size in 2D and 3D models can resolve flow eddies larger than about

Table 6-1. Model Mesh Typical Characteristics.

Dimensions	Typical applications	Typical sedimentation results	Equations averaging*	Typical mesh resolution in areas of interest: Longitudinal or flow direction	lateral or cross-flow direction	Vertical
1-Dimensional Longitudinal	Rivers, canals	Sediment loads, deposition/erosion by channel reaches, dredging volumes	Over cross section	1 to 5 waterway widths plus natural or structural flow controls, dependent on stream size, slope, and uniformity of shape	None–Averaged	None–Averaged
2-Dimensional Horizontal	Rivers and unstratified lakes, estuaries, bays and inlets	Depth-averaged suspended sediment concentration variation in horizontal directions, deposition/erosion along and across waterway	Over depth	Tens to hundreds of feet	Tens to hundreds of feet with minimum of 10 computation points	None-Averaged

(Continued)

Table 6-1. (*Continued*).

Dimensions	Typical applications	Typical sedimentation results	Equations averaging*	Typical mesh resolution in areas of interest: Longitudinal or flow direction	Typical mesh resolution in areas of interest: lateral or cross-flow direction	Typical mesh resolution in areas of interest: Vertical
2-Dimensional Vertical	Stratified waterways with simple cross-section	Vertical distribution of suspended sediment, mobile sediment beds	Over width	Tens to hundreds of feet	None-averaged	Order of feet with minimum 4 to 20 computation points depending on depth
Three-dimensional Hydrostatic	Waterways with significant secondary currents, stratified waterways with complex geometry	Suspended sediment concentration in three dimensions, significant flocculation of fines, mobile sediment beds	None	Tens to hundreds of feet	Tens to hundreds of feet with minimum of 10 computation points	Order of feet with minimum 4 to 20 computation points depending on depth
Three-dimensional Nonhydrostatic	Zones of waterways where vertical accelerations are significant	Usually erosion around structures	None	Inches to feet	Inches to feet	Inches to feet

**Models integrate, average, or approximate over individual mesh cells.*

200 m, but smaller eddies and their transport effects will be lost. A mesh with horizontal cell dimensions about the same as the navigation channel width will not be able to reproduce the physics of transport processes within the navigation channel. A higher resolution can capture finer details but requires more model run time and may introduce other problems such as instabilities. Appendix B illustrates how these factors have been resolved in some past model studies.

Model mesh water boundaries are typically located at flow controls, such as dams, near gauging stations where reliable data are available, or, in the case of ocean boundaries, far from the entrance channel at a location where tidal constituent and wave data are available or can be synthesized.

As with good hydrodynamic modeling, final model mesh resolution and boundary condition specifications should be determined by sensitivity testing – (1) does refining or coarsening of the mesh change the results, and (2) will changes (such as improvement plans) within the problem area affect the boundary area. Both are addressed in the next section.

Mesh resolution is also discussed in the "Sensitivity Testing" section in the following.

Boundary Conditions

Typical open boundary hydrodynamic conditions consist of discharge or flow velocity specified at inflow boundaries and water level or a wave radiation condition at outflow boundaries. Accompanying inflow sediment boundary conditions typically consist of sediment load in some form – bedload and/or suspended load by concentration – and characterization of those loads by grain size or settling characteristics. Outflow boundary conditions may be specified in the same way, but many models use the computed sediment load at the nearest nonspecified cell combined with the computed outflow to compute an outflow sediment rate, that is, the zero-gradient concentration condition.

Some sediment model programs offer an equilibrium noncohesive load option for upstream boundaries – the program calculates noncohesive transport capacity from the specified hydrodynamic conditions and imposes that load as the noncohesive boundary condition. This approach does not provide a reliable wash load boundary condition and wash load may deposit in a downstream reach. In the absence of an equilibrium load option, an upstream supply reach – a feeder section that will erode and supply the next reach with an appropriate noncohesive sediment load if the incoming boundary supply is too low and will allow deposition of the incoming supply is too high – can be employed. Hall et al. (2017) describe both approaches. Note that these options can be employed together as well, that is, an equilibrium boundary can feed into a supply reach.

Ideal inflowing sediment load boundary specifications come from gauged locations with a long data record. When such data are not available, a good estimate should be obtained. For inland and estuarine channels, the next best estimate is from a watershed model as used in the Tenn–Tom Waterway and Atchafalaya Bay studies described in Appendix B. As a last choice when schedule and budget do not permit the aforementioned choices, data for the boundary specification can be developed by desktop analyses such as the proportional sediment yield calculation described in the Cape Fear River of Appendix B. For coastal inlet and offshore channels, boundary conditions may be a shoreline model that considers longshore and cross-shore transport, as described in Appendix B for the Mouth of Columbia River, Cape Fear River, and Long Island.

Initial Conditions

Initial sediment conditions for a model simulation include depth and characteristics of sediment in the bed, including layering, and suspended sediment concentrations and size (or settling velocity) distributions everywhere in the mesh.

Necessary sediment characteristics are those appropriate for the type of bed sediment, as discussed in Chapter 2 and Chapter 5, "Field Observations." Data for these specifications can be obtained from analyzed field samples; however, samples are often sparse and taken under conditions that may no longer be representative of the bed. For that reason and others, a spin-up simulation period is required to establish quasi-equilibrium conditions among the forcing functions, transport rates, and bed elevation and characteristics.

The sediment bed represents the history of the waterway, and therefore, initializing it precisely would require extensive field data or a very long model simulation (on the order of decades) with detailed boundary condition information to replicate the spatial characteristics of the strata thicknesses and composition. This level of model fidelity is impractical. Hence, it is necessary to recognize that the sediment bed initialization in the model will necessarily represent an approximation of the true sediment bed, and there will be errors in the model results associated with this approximation that should be considered when validating the model and interpreting the results.

Spin-Up Simulation

Once the initial sediment bed has been approximated, it is usually necessary to perform a spin-up simulation. The primary purpose of this spin-up simulation is to allow the sediment model to "sort" the surficial bed material, such that high-energy locations are armored (i.e., the more erodible classes are winnowed from the bed, preferentially leaving less erodible classes in the surface layer) and lower-energy locations are surfaced with suitable classes.

It is preferable to perform this sediment spin-up in such a way that sediment can be eroded and deposited from the bed, but the bed elevation does not change. That is, spin-up should be decoupled from morphologic change. This is to ensure that the large variations in sediment concentration associated with the initial bed sorting process do not induce large variations in bed morphology. Each sediment model will have different means of imposing this condition on the solution, but most have this capability.

Spin up should be performed using hydrodynamic conditions that are typical of the system. For example, a model of a river can have a spin-up simulation performed with a typical hydrograph. A model of an estuary could have a spin-up performed over a typical series of seasons.

The duration of spin-up required can be loosely defined as the time required for large changes in the sediment bed adjustment to occur. For models of rivers, this can be ascertained by measuring the steady-state sediment mass flux at the inflow and outflow, and determining the time required for them to achieve approximately the same values (i.e., inflow approximately equal to outflow). For estuaries and coasts, the required spin-up time can be approximated by observing the sediment concentrations at various locations and determining when the variation in concentration as a function of the relevant forcing mechanism (e.g., wind waves) achieves a consistent relationship.

The spin-up time can be relatively short in unstratified hydrodynamic simulations, but much longer in stratified and sedimentation simulations. The bed level and composition, transport rates, and hydrodynamic response to initial bed change must all come into quasi-equilibrium to produce a valid set of restart conditions for validation and test results. Spin-up times cited in the literature range from 30 days in a river reach simulation (Hall et al. 2017) to one year in a large estuarine system (Dynamic Solutions 2010). Appendix B provides spin-up times for several model studies.

Time Steps

Numerical models progress in time by taking discrete time steps. Because time is a continuum, this discretization introduces errors in model accuracy, and potentially in model stability.

The time-stepping of numerical models is typically divided into explicit and implicit time discretization methods. Explicit methods rely on information at the previous time step to solve for information at the present time step. These result in computationally efficient solutions of the model equations, because each equation can be solved independently of the others (i.e., there is no parallel system of equations to solve). However, explicit methods are also subject to stability criteria, associated with the time required for information to pass through one computational unit (i.e., cell, element, volume, and so on). This stability criterion, termed the Courant criterion, is a function of the model resolution and time step, expressed as the Courant number, *Cou*:

$$Cou = \sum_i \frac{u_i \Delta t}{\Delta x_i} \tag{6-1}$$

where

u_i = Flow velocity in the *i* direction,
Δt = Time-step length,
Δx_i = Mesh size in the *i* direction, and
$i = 1$ to the number of dimensions being modeled.

Hence, the more refined the digital model mesh is, the shorter the time step must be to satisfy the Courant limitation of $Cou < 1$.

Implicit methods rely on information at the present time step to solve for other information at the same time step. These methods are much less computationally efficient than explicit methods, because the entire system of equations for the entire mesh must be solved simultaneously. However, implicit methods are not subject to Courant constraints. The only significant limitation on the time step for implicit methods is the characteristic frequency of the phenomena to be simulated. For example, a hydrodynamic model of an estuary with an implicit scheme will require time-step lengths that are sufficient to describe the tidal variation, which has a typical period on the order of hours. Therefore, the time steps can be anywhere from 10 to 60 min in length, depending on what is necessary to approximate the tidal signal.

When sediment transport is added to a model, and especially if multiple grain classes are simulated, fully implicit solutions are typically impractical with presently available computers. A fully implicit solution of sediment transport equations, even if they were decoupled from the hydrodynamics, would require the simultaneous solution of equations for suspended sediment, bedload sediment, bed composition, and the bed elevation. For a model simulating four grain classes, this would result in 13 simultaneous equations per computational point times the total number of computational points in the mesh – a large computation burden. Because of this, most sediment transport equation solvers rely on explicit methods, and/or on an alternative method: the operator-split method. Multiple operator splitting methods are available to be employed by models, but each of them relies on the same basic principle. The model equations are "split" into spatial components (advection and diffusion) and temporal components (storage and source/sink). Each of these is solved separately, and then the results are combined to form the solution.

These solution methods have the great advantage of attaching all the sediment bed exchange terms to the temporal solution, which is a local solution at each computational point. These local solutions can then be solved either implicitly or explicitly, depending on the model (e.g., from the previous example, 13 simultaneous equations are not a significant computational burden for just one computational point). However, these methods also have time-step constraints. Any finite operator split necessarily introduces error, and therefore, the time-step constraint in this case is a practical constraint. The objective is to find the maximum time step that can be used in the model without introducing error into the solution that exceeds an acceptable tolerance. In this respect, it is directly analogous to the mesh convergence methods discussed in the "Sensitivity Testing" section.

If an explicit formulation is employed, the Courant criteria will, in general, be the controlling criteria for determining the permissible time-step length. If an operator-split method is being employed, multiple methods can be used to develop an initial estimate of the time step, but the final selection of the time step can be ascertained through the use of a systematic sensitivity analysis, analogous to that employed for mesh convergence testing.

W. A. (Tony) Thomas of Mobile Boundary Hydraulics, LLC, developed a procedure for selecting a time step in 1D sediment models and reported successfully applying it in 2DH models (personal communication, April 18, 2022). It consists of these steps (Gee 1982):

1. Select three representative steady-state hydrodynamic conditions, such as low, medium, and high flows, to drive the sediment model. In estuarine waters, a steady state may consist of a single, repetitive tide with constant inflows. In a wave-dominated environment, it may consist of a repetitive monochromatic wave.
2. For each hydrodynamic condition, select a short starting time step, such as the same one used by the hydrodynamic model, or one equal to depth divided by sediment settling velocity, and run a spin-up simulation with the aforementioned boundary conditions. Save the ending bed conditions for use as the initial conditions of the next step. Record and plot the time history of modeled bed change at the location where bed change is the largest.
3. Increase the time-step length and run the model for a few time steps or up to half the spin-up duration. In early stages, doubling the time step is suggested. As the tests proceed, smaller multiples may be better. Save the ending bed conditions for use as the initial conditions

of the next length of step to be tested. Record and plot modeled bed change time history at the location where bed change is the largest.

4. Repeat Step 3 until the results become unstable or unreasonable. The last time step length at which bed elevation is approximately stable and sediment inflow approximately equals outflow from the cell with maximum change is an appropriate choice for that hydrodynamic condition.

Another procedure consists of these steps:

1. Perform a simulation with a large time step. For implicit models, this may correspond to the time step used for simulation of the hydrodynamics. For some models (such as 1D models with explicit implementations, where the Courant limitation is not a significant constraint because the spatial resolution is the distance between successive cross sections), the initial time step selected may be much larger than that required for the hydrodynamics.
2. Perform the same simulation with successively smaller time steps (e.g., reduce each time step length by 1/2 in each succeeding trial.)
3. If reducing the time step to 1/4 of the hydrodynamic time step does not alter the results sufficiently to warrant that additional level of refinement, try increasing the time step in increments until the results are sufficiently altered to indicate a problem.
4. Select the largest time step for which further change in the time-step size does not alter the results significantly.

Regardless of the time step selected through the systematic procedure described previously, the selected time step is also constrained by the need to resolve sediment processes of interest to the specific modeling effort. Site-specific sedimentation process considerations may include

- Flocculation of cohesive sediment,
- Suspended sediment transport,
- Bedload or density flow transport,
- Bed exchanges,
- Bed consolidation/fluidization, and
- Bed armoring.

The combination of mesh resolution, time-step size, time period modeled, and computer hardware dictate the clock time required to complete a simulation. A simulation producing results only after running for days may adversely affect project schedules.

The morphologic time step – a multiple of the hydrodynamic and transport time step – has been implemented in some sedimentation models to reduce simulation time. It consists of simulating a specific period in normal time-stepping mode, that is, with both hydrodynamic and sediment calculations at every step, then extrapolating the calculated bed change forward for a morphological step of length Δt_{morph}, given by

$$\Delta t_{\text{morph}} = f_{\text{morph}} \Delta t_{\text{hydrosed}} \tag{6-2}$$

where f_{morph} is the acceleration factor, and $\Delta t_{\text{hydrosed}}$ is the hydrodynamic and sediment transport time step. Bed changes during the morphological time step are calculated by either Equations (6-3a) or (6-3b)

$$\Delta h_{\text{morph}} = f_{\text{morph}} \Delta h_{\text{hydrosed}} \tag{6-3a}$$

$$\Delta M_{\text{morph}} = f_{\text{morph}} \Delta M_{\text{hydrosed}} \tag{6-3b}$$

where

Δh_{morph} = Bed elevation change during a morphological time step,
$\Delta h_{\text{hydrosed}}$ = Bed elevation change during one time step of hydrodynamic and sediment transport modeling,
ΔM_{morph} = Net bed mass exchange during a morphological time step, and
$\Delta M_{\text{hydrosed}}$ = Net bed mass exchange during one time step of hydrodynamic and sediment transport modeling,

Use of a morphological time step relies on two assumptions: (1) Sediment system is trending toward morphologic equilibrium with the hydrodynamics and the sediment load, and therefore, the morphology can be "accelerated" to make this process happen more quickly in the model, and (2) updated hydrodynamics using the accelerated morphological change can correct any tendency to overshoot the actual solution.

Users of a morphological time step should test this latter assumption, which means, for example, that excess deposition in an accelerated step will be eroded in the next calculated step, thus correcting the overshoot. Overshooting bed changes may also lead to numerical instabilities, so a careful inspection of results is necessary to prevent spurious results. Morphologic time-stepping can be very useful, but must be used with extreme caution, because an upper limit on the morphologic time step, like that for the regular time step, depends on the modeled site's hydrodynamic and sediment characteristics. Examples of morphological time-stepping applications include the following:

- Donnell et al. (1991) employed a 1 h transport time-step transport model to run one-year continuous simulations, each containing specific events, of mixed sediment discharges into Atchafalaya Bay, Louisiana, then extrapolated [Equation (6-3a)] those bed changes for

10- and 15-year increments before rerunning hydrodynamics and transport for a total simulation of 50 years. The Atchafalaya Bay section of Appendix B shows how the predicted delta areal extent was confirmed by observed conditions.
- Lesser et al. (2004) modeled noncohesive sediment under tidal and wave-driven forcings for laboratory experiments and harbor entrance at Ijmuiden, The Netherlands, and concluded that morphological time steps [Equation (6-3b)] 20 to 100 times the hydrodynamic steps (order of seconds to minutes) produced acceptable results for those conditions.
- Brown et al. (2019) employed an acceleration technique similar to Equation (6-3a), derived from a modification of the bed porosity. An acceleration factor of 10 was applied to simulate 50 years of West Bay, Louisiana, delta growth associated with sediment diversions.
- Meselhe et al. (2021) used an acceleration factor [Equation (6-3b)] of 60 to simulate sedimentation processes at the Wax Lake Outlet in Atchafalaya Bay, Louisiana.
- Mathew and Winterwerp (2022) tested acceleration factors [Equation (6-3b)] ranging from 1 to 15 in modeling Passaic River fine sediment transport. They concluded that a maximum acceleration factor of 7 was appropriate for depositional periods, but only a factor of 1 gave acceptable results for erosional periods because of varying bed erosion resistance with depth. They also found that accelerating bed mass exchange with Equation (6-3b) produced more reliable results than accelerating bed elevation change with Equation (6-3a).

Best practice for the use of a morphologic acceleration methodology is to test the selected acceleration factor against a base simulation without any acceleration and compare the results. This is a necessary, but not sufficient, condition for use of these methods. Even if the selected time acceleration factor passes this test, the modeler must be aware of the assumptions employed in the specific acceleration techniques used by their selected model and interpret the results in light of those assumptions.

Model Parameters

Sediment transport models employ many adjustable parameters and modelers may overtune them, producing results that appear reasonable for the wrong reasons. Model user manuals often recommend standard ranges for each parameter and users must exercise caution when assigning values outside those norms. Sensitivity testing, as described below, can be used to inform modelers on the effect of their choices.

Sensitivity Testing

Sensitivity testing serves multiple purposes in sedimentation modeling, including establishing sufficient mesh resolution (convergence), suitable boundary locations, model coefficients, and an appropriate time step. Sensitivity testing can also be used to estimate model uncertainty bounds in the absence of sufficient observed data, as discussed in Chapter 5.

Any finite discretization of a continuous domain will result in some dependence of the solution on the resolution of the discretized domain. That is, any refinement of the mesh will necessarily result in some change in the model result, indicating that the result is dependent on the resolution. This dependence (i.e., the degree to which the result changes with refinement) is defined as the level of mesh convergence.

Mesh resolution testing is the process by which the modeler determines the level of spatial resolution necessary for the solution to be acceptable. That is, what resolution is necessary for the model resolution to have negligible influence on the results.

This determination is often qualitative and practical in nature. The basic procedure is as follows:

1. Perform a simulation with a coarse mesh.
2. Perform the same simulation with successively finer meshes.
3. Select the coarsest resolution for which further refinement does not alter the results sufficiently to warrant that additional level of refinement.

Because errors tend to asymptotically approach zero as meshes are further refined, it is usually not difficult to select the desired level of resolution – no finer than necessary to obtain consistent results. A finer resolution mesh requires more computer time and may need different turbulent diffusion coefficients.

Models with dynamically adaptive mesh resolution change the mesh as needed to reduce user-specified error but do not eliminate the need for mesh resolution testing. For adaptive meshes, the same procedure as mentioned previously is performed, except that the specified error tolerances are successively reduced to result in progressively more refined adaptive meshes.

One thing to consider carefully when performing mesh resolution testing is that the tests will work properly only if the underlying physics are properly defined. Specifically, turbulent mixing coefficients (e.g., eddy viscosities, turbulent diffusion) are physics-based properties and should be specified as such. Unfortunately, they are sometimes misused as tuning parameters for stabilization or validation. If the mixing coefficients are too large, mesh convergence will not reveal the proper resolution for the model, as discussed subsequently.

For proper model development, turbulent mixing coefficients should be constrained by their intended function. Turbulent mixing coefficients arise from the Reynolds decomposition of the Navier–Stokes equations, whereby the velocity is separated into a steady component and a time-varying component. The temporal scale of this separation (i.e., the time interval over which the equation is integrated) is a function of the temporal scales to be resolved in the model. Because temporal and spatial turbulence scales are, in general, closely linked, it means that the choice of temporal scale to be resolved influences the choice of spatial model resolution.

In general, mesh resolution should be sufficient to resolve mixing phenomena that are intended to be directly simulated by the model (e.g., large eddies), whereas the energy dissipation and/or constituent diffusion associated with phenomena at smaller temporal and spatial scales should be captured by the turbulent mixing coefficients. If the selected coefficients are too large, they will overwhelm the mixing at larger scales, and the additional resolution intended to directly simulate the larger-scale phenomena will prove useless. This problem is best understood at the limit. For example, the specification of a near-infinite turbulent diffusion coefficient will result in a spatially uniform concentration. In this case, a very coarse mesh and a very fine mesh will yield identical results, and no insight is gained by performing mesh resolution testing.

Careful mesh resolution testing is a fundamental aspect of good numerical modeling practice. Failing to do so can yield a model for which the results are heavily dependent on mesh resolution and not on the underlying physics.

Experienced modelers have a general idea of how far boundaries should be from the problem area to allow the problem area to be isolated from the boundary condition. However, sensitivity to boundary condition locations should be tested at the beginning of a numerical model application and the issue should be periodically revisited. The issue can be illustrated with an exaggerated problem.

> Suppose a river reach model has a gauged discharge boundary a few miles upstream and a gauged stage boundary a few miles downstream. Model simulations of a proposed bridge with multiple piers show it causes a hydraulic choke in the reach, i.e., an upstream backwater is created and supercritical flow to a hydraulic jump occurs below the bridge. Those extreme changes will profoundly affect sediment transport in the reach and those changes might propagate throughout the modeled area, but if the upstream boundaries have specified sediment discharges, they cannot change and the results are contaminated.

Although this example is exaggerated, it illustrates the potential for conflict between modeled conditions in the problem reach and at the boundaries.

The standard sensitivity method to test boundary locations is to run a base simulation with selected boundaries and then move the boundaries closer or further away and rerun the simulation. If the closer boundaries are sufficiently far away, there should be at most minor differences in model results over the problem reach. Note that this sensitivity test does not absolutely prove the reverse–severe changes to the model in the problem reach might still affect the boundaries, as described in the exaggerated problem mentioned previously. Testing of different boundary locations can be thought of as a necessary but not sufficient test. The modeler must be alert to any plan changes – channel enlargements, training structures, etc. – that might impact the boundaries.

Validation

The first requirement for successful sedimentation modeling is a well-validated hydrodynamic model. See Brunner et al. (2020) and Wang et al. (2008) for procedures in hydrodynamic model validation. The sediment model can then be validated with confidence that the hydrodynamics are correct.

Ideally, every navigation channel sedimentation model is tested against accurate field (prototype) observations of deposition and erosion to demonstrate that it is capable of producing useful predictions – those that can be used to make sound judgments about channel sedimentation. Further, the range of prototype conditions used in validation should encompass the entire range of conditions and durations to be tested in the model. For example, a model validated only for low flow conditions is not necessarily validated for high-flow conditions. This ideal may be compromised by poor or incomplete prototype data. If useful protype data are limited or unavailable, validation will be limited or absent and model uncertainty bounds must reflect those limitations, as discussed in the "Interpreting and Reporting" section of this chapter.

Many modelers employ a two-step terminology for the validation process. In step one, the model is "calibrated" against site-specific observations on key measures of interest (e.g., deposition rates) over a historical period. Adjustable model parameters (e.g., boundary roughness and grain sizes) are tuned until there is good agreement between the model and the historical data. In step two, model "validation," the model is a rerun for a different time period without changing model parameters. If the agreement is sufficient, the calibrated model is said to be validated. We believe that the two-step terminology is inadequate for good modeling practice.

When the two-step calibration–validation terminology came into use in the 1970s, it was opposed by the US Army Corps of Engineers Waterways Experiment Station (WES) Hydraulics Laboratory, which had long used a different terminology. In that lexicon, instruments such as flow meters were "calibrated" but numerical and physical models were "verified" to field data. Verification was defined as a multistep process involving one to many sets of field data (Simmons 1977). The WES objection expressed a concern that stopping at two steps sometimes produced an inferior model giving a false sense of certainty. WES asserted that the process should be iterative among all available data sets until the best overall agreement with field observations was achieved. Multiple data sets under a wide range of conditions were strongly encouraged, because many systems exhibit characteristics unique to the circumstances. For example, boundary roughness coefficients and grain size distributions often change with variation in flow or wave conditions, producing hysteresis in stage–discharge–wave–sediment load relationships, as described in Chapter 2. Trying to reproduce all of them with a single set of adjustment parameters may produce misleading results when conditions change.

In this document, we use the term validation for the process of using one to many observed data sets to demonstrate that a sedimentation model is useful for its intended purpose. The validation data sets should encompass the conditions to be tested, for example, a range of low to high flow velocities, waves, and or sediment supply, depending on site conditions. A comparison of model results with two or more unrepresentative prototype conditions does not produce a validated model. Time and funding resource limitations often constrain the validation process to too few prototype data sets. The proper response to resource limitation is a careful use of model uncertainty bounds.

The validation process should produce well-defined uncertainty bounds to be considered in the interpretation of model results. Chapter 5 section "Uncertainty in Predictions" provides methods for estimating uncertainty bounds, including those imposed by poor or limited prototype data.

The phrase "useful for its intended purpose" means that the model results can be interpreted by knowledgeable sedimentation experts to render sound engineering judgments for the modeling objectives. See the following "Fitness for Use" section for additional information.

An important principle produced by the "Epistemic and Structural Uncertainties" discussed in Chapter 5 might be called the modeler's version of the Heisenberg uncertainty principle. Investigators can specify how good a sedimentation model's validation will be or how much validation will cost (time and money) but not both. In other words, specifying quantitative validation quality – goodness of fit to observed data – may doom a modeling effort to failure. The better practice is to produce the best validation possible within schedule and budget constraints and use the validation results to set confidence limits on the model results, as discussed in the "Interpreting and Reporting" section. Doing so empowers decision makers to properly account for uncertainty and risk in their selection of plans.

Statistical measures, discussed in Chapter 5 "Uncertainty in Predictions," are an essential component of model validation; however, they can mislead under some circumstances. For example, a model time series of sediment concentrations with proper magnitude and time variation will show a very poor RMSE error statistic if they are out of phase with the observed data. Graphical displays should be an integral part of any validation effort. Correlation coefficients (coefficient of determination) can be misleading for several reasons, including systematic error blindness (e.g., Willmot 1981).

Sedimentation model validation works best when it employs a weight of evidence approach that includes a visual inspection of model results plotted with observed time series and spatial data and evaluation of model performance statistics. The weight of evidence approach recognizes that the model only approximates complex sedimentation processes, and perfect agreement with observed data, itself marked by errors, is impossible. Instead, the goal is general agreement that provides the user with confidence to interpret and report model results in light of established uncertainties.

Testing

The testing, or production, modeling phase produces quantitative results that a sedimentation processes expert interprets for management, design, or regulatory purposes. Direct use of precise model predictions without interpretation is discouraged.

Test conditions should be chosen to best represent those the project will experience; however, they may need to include those conditions that stress the project. As noted in Chapter 2, episodic events – floods and storms – often cause greater sedimentation events than months of normal conditions. The conceptual site model described in the "Problem Definition" section will inform the decision on what conditions should be included in the validation and production testing programs.

As with validation simulations, production testing results must be taken after a spin-up period or from a hotstart condition. A basic testing rule is that fewer changes to existing conditions makes interpreting test results simpler and more reliable. For example, a simulation with one training dike installed provides results that are plainly the effect of that one dike; whereas, a simulation with many planned elements – dikes, dredging, channel realignment, and others. – cannot distinguish the individual effects of each element. The time and cost of testing must be balanced against the need for evaluating separate elements of proposed changes.

The question of how long to simulate a changed condition, such as planned construction of a dike field, is similar to that of a sediment model spin-up. It is complicated by the fact that natural waterways never reach true equilibrium; they respond to seasonal and annual cycles plus long-term changes in hydrodynamic and sediment conditions, and at most, approach a quasi-equilibrium over engineering timescales. The key question then arises: What length of simulation time is sufficient to obtain model results that reliably demonstrate the effects of proposed changed conditions over the time periods specified in the objectives?

The length of time for a sediment model to reach quasi-equilibrium from a plan condition depends on waterway flow and sediment characteristics plus the magnitude of the change. Some examples are as follows:

- Schoellhamer et al. (2008) found that perturbations to San Francisco Bay models (2D, 1D, and 0D) for cohesive sediment required 14 years to return to equilibrium after a 10% increase in water depths. A 10% overall depth increase over San Francisco Bay is equivalent to removing hundreds of millions of cubic meters of sediment. That plus the slow response time suggests that sediment supply is a limiting factor there.
- Pittaluga et al. (2014) tested a 1D river model of a 6.4 km reach of the Magra River and found that morphologic recovery after a flood required 1 year to achieve approximate equilibrium and 3 years to reach exact equilibrium sediment discharge through the reach.
- Cooper (1993) analyzed bathymetric data from the river-dominated Mgeni Estuary and concluded that approximate equilibrium occurred within a few months after major floods eroded sediment from the estuary, but a total return to former morphology took about 70 years.
- Little and Biedenharn (2014) used a $\pm$ 0.8% change in channel volume (below top bank) criterion for "dynamic equilibrium" in a geomorphic analysis of Mississippi River reaches.

No simple procedure exists to determine the appropriate quasi-equilibrium time, so the modeler must determine the appropriate simulation durations by trial and ensure that the model results represent a depiction of the plan effects appropriate to the study objectives and fitness for use.

As discussed in the Chapter 5 section "Uncertainty in Predictions," predictions of actual future values, such as deposition/erosion rates, are affected by the model's calculated uncertainty plus aleatory uncertainty – future water levels, flows, waves, and sediment supplies will be time-varying with random components (i.e., stochastic), which are predictable only probabilistically at best. Deterministic future sedimentation quantitative predications must be expressed in probabilistic terms based on the joint probabilities of future conditions (e.g., Ochi 1990).

A better alternative to predicting future quantitative sedimentation values is base-to-plan testing – comparing one model result (e.g., existing, or base, conditions) to another (proposed, or plan, conditions) as relative changes. That is the highest and best use of model results. Such comparisons reduce model knowledge and structure errors, because they are embedded in both base and plan model results, and they minimize future-casting disparities caused by stochastic boundary conditions. They can be expressed as percentage or ratio changes in sedimentation rates with specified uncertainty bounds, as discussed in the "Validation" section. One rule of thumb is that the uncertainty bounds on base-to-plan percentage comparisons are about half those on predictions of absolute, unit-based results, as discussed in Chapter 5.

Record Keeping

Meticulous recording of every step in the modeling process provides essential continuity to the process, prevents duplication of effort, makes report writing easier, allows better quality control checks, and serves as a project record. Records should be in plain language so that another modeler can check or continue the work if necessary. Filename conventions should be described in a comprehensive metadata file and faithfully adhered to. The details of each model run should be documented in either a log book or by adding notes to the modeling files or both. We suggest using both, as the log book has more room for project details, whereas the modeling notes are part of the electronic project record. The model run documentation should include the run date, time, name of the run, the run's relationship to previous runs, what has been changed, why it was changed, and the results of the run. Some navigation studies last for years, therefore, it is critical that the version number of the software is associated with each run. The log book should also contain details of the data that were used for the project, including the source of the data, type of data, whether the data are in raw form or processed data, and the datum plane(s) used. Finally, the log book should include documentation of the steps taken to adhere to the quality assurance/quality control (QA/QC) plan.

The log book can be either a literal book or a digital text or spreadsheet file. Physical logbooks should be protected from damage by fire, water, and vermin. Digital logbook files, such as model input and output files, must be backed up frequently to both onsite and offsite locations.

Fitness for Use

The "Problem Definition" section at the beginning of this chapter noted the significance of quality issues – fitness for the intended use. The previous sections describe how model results quality may be improved through rigorous digital model development, sensitivity testing, validation, and testing. Rigor must be matched to the client's quality requirements and to available resources.

Matching rigor to requirements means that sensitivity and validation results should be at least sufficient to the needed quality. Scientific certainty is often beyond reach within available resources. Achieving model results that are "good enough" to meet the need is the minimum goal. The gap between good enough and perfection is defined by uncertainty bounds as defined in Chapter 5 and must be explicitly reported, as discussed in the following section.

INTERPRETING AND REPORTING

W. A. (Tony) Thomas, creator of the HEC-6 and HEC-6T sediment transport models, often said, "The purpose of modeling is insight, not answers." The late Robert G. Dean, renowned coastal engineer, said, "The best modeler is a skeptical modeler."

Thomas's and Dean's statements are true. Convincing clients of that truth is hard; clients want numbers. Therefore, interpreting results carefully is among modelers' most delicate and important tasks. Sediment transport models produce answers – actual numbers about concentration, deposition, and erosion. Sedimentation experts must examine those numbers in light of their knowledge and experience to gain proper insights so that they can provide the quantitative or qualitative answers that clients need and the results justify. They can accomplish this goal by

- Restricting practice to areas of their personal expertise.
- Being skeptical of model results – are they physically plausible, reasonable, and consistent with expectations? If not, can deviations be explained as model error or unanticipated physical responses? Such explanations require highly qualified expert knowledge of modeling and sedimentation processes.
- Applying appropriate uncertainty bounds and employing only significant digits in results. [Always conform to significant digits rules (Wikipedia 2023) in presenting results unless more digits are needed to illustrate small differences.]
- Using Base-to-Plan (model versus model) comparisons to maximum extent possible.
- Explaining how the numerical results should be interpreted, including their limitations.
- Ensuring that conclusions satisfy objectives without overstating the case.

Another important principle derives from this description of interpretations and the previous discussion of model validation issues – the most important qualification for a sedimentation modeler is a thorough understanding of the hydrodynamic and sedimentation processes of the modeled site so that the modeler readily recognizes physically implausible data and properly interprets all results. A knowledge of numerical modeling is necessary, but it is not sufficient.

Reporting model results should follow the form of an excellent lab report as taught in engineering classes – at a minimum a clear and complete statement of objectives, approach, site description, field observations details, model description (with limitations explained), validation with uncertainty explained, results with discussion, and conclusions.

Every model study report must explicitly state the uncertainty bounds on results – on every table and graphical displays of numerical results. Tables and figures will almost certainly be extracted from the report and used elsewhere, without any reference to the uncertainties stated in the text, unless the authors make it difficult to do so. Avoiding misuse of model results is a fundamental responsibility of the modeler (NAP 2012).

MONITORING AND FEEDBACK

Completion of a sediment modeling project often produces sighs of relief and an urge to get on with the next one; however, some retrospective considerations are worthwhile.

- Whenever possible, monitor project performance and evaluate modeling methods.
- Identify mistakes to improve future efforts.
- Provide knowledge on processes and methods to the profession.

As obvious as these actions sound, they prove difficult in practice. First, as-built projects seldom reproduce exactly the modeled conditions. Second, clients are seldom interested in spending additional money on projects if they performed more or less as planned. Finally, new projects are usually more alluring than completed ones.

MANAGING MODEL STUDIES

The management of numerical modeling studies has its own particular challenges to bring a project to a successful conclusion. Once a project is initialized by funding authorization, work may proceed differently depending on the client – federal, state, or local agency, university, or nonprofit group – and modelers' employer. Terms may differ as well. The term *project manager* is used here in the general sense of supervising the project, including modeling. In some organizations, the project manager has a different role to play.

Private consultants generally receive or identify a scope of work or statement of work (SOW) from potential clients and then respond by preparing either a technical and cost proposal or a qualifications package. A technical and cost proposal should include a numerical modeling approach, cost estimate broken down by subtasks specified in the SOW and project schedule. The next step is to identify the key modeling team members. The lead modeler(s) should prepare a detailed modeling approach that should include identification of the validation targets and the methods that will be used to ensure those targets are met. The modeling approach should also include quality assurance/quality checks and methods that will be used during the project. The modeling approach drives the cost proposal preparation and guides the development of the project schedule. It is recommended that the senior project team members lead the preparation of the cost proposal, and therefore, the entire project team understands how the work will proceed and any subtasks not specified in the SOW are identified and included in the project budget and schedule.

Typical tasks included in a modeling study are identification and analysis of all existing data that will be provided by the client; collection of other available existing data; model setup including mesh (or grid), digital elevation model, and model boundary condition development; hydraulic/hydrodynamic and sediment transport model validation; sensitivity analysis testing; alternatives analysis; status meetings; review of deliverables; and reporting. The project schedule should include enough detail to identify critical path tasks. The project schedule should include enough buffer between tasks to allow for internal and client reviews and time for subtasks requiring additional unplanned tasks to be completed. For example, it is often the case that data provided by the client or via external sources is in a raw form that requires extensive processing to ensure sufficient quality, quantity, and in the correct format (with datum) for use in the modeling study.

The project manager should set the project up in an administrative tracking system to allow monitoring of the project schedule and internal billing to the project. An internal kickoff meeting should be held to establish project responsibilities and lines of communication for the project. Each modeling team member needs to understand the project's schedule and budget for their particular task/subtask, QA/QC requirements and when project deliverables are due to the client. Internal meetings should be held in preparation for client recurring meetings for the duration of the project to evaluate progress and resolve any challenges that impact cost or schedule. Meetings (including video conferences) with the client should be conducted to demonstrate progress, discuss any data or modeling issues that may impact cost or schedule, and manage client expectations. Detailed meeting minutes should be taken during the meetings and submitted to the client as part of progress reporting.

Management should create and enforce a data backup system using redundant onsite and offsite storage locations. All project files, including correspondence, field data, analyses, and model files must be backed up at least daily.

One of the most challenging steps in managing numerical modeling studies is monitoring progress. A passive management approach is not effective for numerical modeling studies, as there is a tendency for modelers to quietly continue to work away at a problem when things do not go as planned. This can quickly put the budget and schedule at risk of overruns. Monitoring can be difficult as the work products are often not readily accessible outside of modeling software. Further, staff tend to want to put the best light on the work by noting that progress is being made. Therefore, the most successful approach to managing numerical studies is to break the work into small subtasks and compare each completed component to the project schedule. This must be done every week. This may be as simple as having another modeler on the team to review the work progress or as complex as multidisciplined all-hands reviews of all the modeling assumptions, input files, and modeling result for major deliverables. In no case should there be only one set of eyes on final products. Reviews must be held frequently. This allows the team to identify problems early and mitigate challenges before time and funds are wasted.

Instituting QA – the processes in place that promote good practices and checks for consistency with objectives and standards throughout a project and QC – inspection of products for conformance to fitness for use criteria are the responsibility of project management. Implementation of QA/QC processes is the responsibility of both management and project personnel, including modelers. Guidance for quality assurance of modeling projects is available from EPA (2002) and guidance for general project QA/QC is given by USACE (2021) regulations and the standard ISO 21500 (ISO 2021).

Written progress reports to the client are essential. They serve not only as part of an effective communication plan, but also prevent later confusion over client decisions. The client point of contact may change during a project, and a well-documented history will inform the new contact person and protect the modeler.

Every modeler should make at least one site visit to see the project environment. A first-hand knowledge of the site and its environment is essential to informed modeling. Features that do not appear on maps – new scoured flow pathways, underwater obstructions, and turbidity patterns – can be discovered. Our personal site inspections have revealed these and more, such as hydraulic structures that were no longer in place and bridges collapsed into the water.

AGENCY MODELING REQUIREMENTS

Large organizations may require specific modeling practices. Some guidance from US Federal organizations is described below. The modeler must be aware of these requirements and others imposed by state or local agencies.

Environmental Protection Agency

The Environmental Protection Agency (USEPA) provides and supports a suite of models for use in environmental studies as part of its Better Assessment Science Integrating Point and Nonpoint Sources (BASINS) toolkit (EPA 2021). Several sediment analysis tools are included, including the 3D numerical model EFDC, mentioned previously.

EPA provides guidance on use of models in environmental studies through its manual, *Guidance on the Development, Evaluation, and Application of Environmental Models* (EPA 2009).

US Army Corps of Engineers

The USACE Hydrology, Hydraulics and Coastal Community of Practice has identified selected engineering model software as preferred for use on Corps of Engineers investigations and has indicated that listed model software should be used whenever appropriate. The preferred model software has been developed by the Hydrologic Engineering Center (HEC) and Engineering Research and Development Center (ERDC). Additional engineering modeling software has been identified as acceptable or allowed for use; the latter list includes legacy software and software developed by others. Both software lists are available to USACE staff for official use only.

For USACE investigations, the selection and application of the model software and the input and output data are subject to organizational quality control and technical reviews (USACE 2021). USACE describes sedimentation modeling practice in discipline-specific engineer manuals, notably

- *Dredging and Dredged Material Management* (USACE 2015a),
- *Deep-Draft Waterways* (USACE 2006),
- *Coastal Engineering Manual* (USACE 2002),
- *Shallow-Draft Waterways* (USACE 1980),
- *Tidal Hydraulics* (USACE 1991), and
- *Sedimentation Investigations of Rivers and Reservoirs* (USACE 1995).

US Department of Interior, Bureau of Reclamation

The Bureau of Reclamation provides and supports a suite of models for sedimentation and river hydraulics, including SRH-2D listed in Table 5-3. Its Technical Service Center offers integrated technical expertise in topics such as flood hydrology and meteorology, sedimentation and river hydraulics, and economic analysis.

REFERENCES

Brown, G. L., J. N. McAlpin, K. C. Pevey, P. V. Luong, et al. 2019. *Mississippi river hydrodynamic and delta management study: Delta management modeling: AdH/SEDLIB multi-dimensional model validation and scenario analysis report*. ERDC/CHL TR-19-2. Vicksburg, MS: US Army Engineer Research and Development Center.

Brunner, G., G. Savant, and R. E. Heath. 2020. *Modeler application guidance for steady vs unsteady, and 1D vs. 2D vs 3D hydraulic modeling*. TD-41. Vicksburg, MS: USACE Research and Development Center, Davis, CA: Hydrologic Engineering Center.

Cooper, J. A. G. 1993. "Sedimentation in a river dominated estuary." *Sedimentology* 40 (5): 979–1017.

Donnell, B. P., J. V. Letter Jr., and A. M. Teeter. 1991. *The Atchafalaya River Delta, Report 11: Two dimensional model*. Tech. Rep. HL-82-15. Vicksburg, MS: USACE Waterways Experiment Station.

Dynamic Solutions. 2010. *Delta EFDC sediment transport model calibration and validation*. Tech. Rep. Sacramento, CA: US Army Corps of Engineers.

EPA (US Environmental Protection Agency). 2002. *Guidance for quality assurance project plans for modeling*. EPA QA/G-5M. Washington, DC: EPA.

EPA. 2009. *Guidance on the development, evaluation, and application of environmental models*. Washington, DC: EPA.

EPA. 2021. *Better assessment science integrating point and non-point sources (BASINS)*. Washington, DC: EPA.

Gee, D. M. 1982. *Guidelines for the calibration and application of computer program*. HEC-6. Davis, CA: USACE, Hydrologic Engineering Center.

Hall, B. M., R. A. Gaines, Y. Lin, and K. H. Waller. 2017. *Merriweather-Cherokee potamology study*. Mississippi Valley Division MRG&P Rep. No. 9. Vicksburg, MS: USACE, Mississippi Valley Division.

Hudson, R. A., F. A. Herrmann Jr., R. A. Sager, R. W. Whalin, et al. 1979. *Coastal hydraulic models*. Special Rep. 5. Vicksburg, MS: USACE Waterways Experiment Station, Coastal Engineering Research Center.

ISO. 2015. *Quality management systems—Fundamentals and vocabulary*. ISO 9000. Geneva: ISO.

ISO. 2021. *Project, programme and portfolio management—Context and concepts.* ISO 21500. Geneva: ISO.

Juran, J. M. 1980. *Quality planning and analysis: From product development through use.* New York: McGraw-Hill.

Lesser, G. R., J. A. Roelvink, J. A. T. M. van Kester, and G. S. Stelling. 2004. "Development and validation of a three-dimensional morphological model." *Coastal Eng.* 51(8-9): 883.

Little, C. D., and D. S. Biedenharn. 2014. *Mississippi River hydrodynamic and delta management study—Geomorphic assessment.* Final Rep. *ERDC/CHL TR-14-5.* Vicksburg, MS: Engineering Research and Development Center.

Martin, J. L., and S. C. McCutcheon. 1999. *Hydrodynamics and transport for water quality modeling.* Boca Raton, FL: CRC Press.

Mathew, R., and J. C. Winterwerp. 2022. "Morphodynamic modeling and morphological upscaling in a fine sediment system." *Adv. Water Resour.* 166 (August): 104224.

Meselhe, E., K. Sadid, and A. Khadka. 2021. "Sediment distribution, retention and morphodynamic analysis of a river-dominated deltaic system." *Water* 13 (10): 1341.

NAP (National Academies Press). 2012. *Assessing the reliability of complex models.* Washington, DC: US National Academies of Science, Engineering and Medicine.

Ochi, M. K. 1990. *Applied probability and stochastic processes: In engineering and physical sciences.* Wiley Series in Probability and Statistics. New York: Wiley.

Pittaluga, B. M., R. Luchi, and G. Seminara. 2014. "On the equilibrium profile of river beds." *J. Geophys. Res. Earth Surf.* 119 (2): 317–332.

Schoellhamer, D. H., N. K. Ganju, P. R. Mineart, and M. A. Lionberger. 2008. "Sensitivity and spin-up times of cohesive sediment transport models used to simulate bathymetric change." *Proc. Mar. Sci.* 9 (1): 463–475.

Simmons, H. B. 1977. *Letter from chief, Hydraulics Laboratory, U.S. Army Engineer Waterways Experiment Station, to Thorndike Saville, technical director.* Washington, DC: US Army Corps of Engineers.

STOWA (Foundation for Applied Water Research). 1999. *Good modelling practice handbook.* STOWA Rep. 99-05. Lelystad, Netherlands: Dutch Dept. of Public Works, Institute for Inland Water Management and Waste Water Treatment Rep. 99.036.

USACE (US Army Corps of Engineers). 1980. *Layout and design of shallow-draft waterways.* EM 1110-2-1611. Washington, DC: USACE.

USACE. 1989. *Sedimentation investigations of rivers and reservoirs.* Engineer Manual 1110-2-4000. Washington, DC: USACE.

USACE. 1991. *Tidal hydraulics.* EM 1110-2-1607. Washington, DC: USACE.

USACE. 1995. *HEC-6 reservoir sediment control applications*. TP-148. Davis, CA: Hydrologic Engineering Center, USACE.

USACE. 2002. *Coastal engineering manual*. Engineer Manual 1110-2-1100. Washington, DC: USACE.

USACE. 2006. *Hydraulic design of deep-draft navigation projects*. Engineer Manual 1110-2-1613. Washington, DC: USACE.

USACE. 2015a. *Dredging and dredged material management*. Engineer Manual 1110-2-5025. Washington, DC: USACE.

USACE. 2015b. *SMART planning feasibility studies: A guide to coordination and engagement with the services*. Washington, DC: USACE.

USACE. 2021. *Quality management*. ER 1110-3-12. Washington, DC: USACE.

Wang, S. Y., P. J. Roache, R. A. Schmalz, Y. Jia, et al. 2008. *Verification and validation of 3D free-surface flow models*. Reston, VA: ASCE.

Willmot, C. J. 1981. "On the validation of models." *Phys. Geogr.* 2 (2): 184–194.

Wikipedia. 2023. "Significant figures." Accessed August 28, 2023. https://en.wikipedia.org/wiki/Significant_figures.

Winterwerp, J. C., T. Van Kessel, B. S. Van Maren, and B. C. Van Prooijen. 2021. *Fine sediment in open water—From fundamentals to modeling*. Singapore: World Scientific.

CHAPTER 7
CONCLUDING REMARKS

Navigation channel sedimentation problems are complex, the product of interacting weather, hydrology, hydraulics, sediment characteristics, and geomorphodynamics, plus human activities. Projects successfully solving those problems require identifying sustainable solutions that may solve the problems and evaluating those solutions to ensure that they satisfy expectations.

The first requirement for a successful solution is a project team with a deep knowledge of channel sedimentation processes typical of the problem area. The sediment dynamics of coastal areas can be very different from that of rivers and cohesive sediment behaviors are very different from those of noncohesive ones. Expertise in one area does not imply expertise in another.

Next, identifying potential sedimentation solutions should be initially expansive, using a systematic evaluation of both usual and unusual solutions. "We've always done it that way" is not a suitable justification to use a standard solution. Following a structured taxonomy like that in Chapter 4, "Solutions to Sedimentation Problems," is advised.

Potential solutions can be evaluated in multiple ways. A combination of field studies, desktop analyses, and modeling in a tiered approach is usually best, but the rigor of the evaluation depends on the stage of the project, resources available, and potential impacts of the project. Adaptive management—monitoring, evaluating, and modifying a project after construction—is a valid approach but should never replace a proper preconstruction evaluation. Trial and error is almost never a good idea.

Evaluating solutions by numerical model has become relatively easy because of standardized numerical model programs and graphical user interfaces. Some engineers fear that it has become too easy. The late Mary

Kendrick, formerly of Hydraulics Research Station, Wallingford, UK, once said, "A numerical model in the hands of an inexperienced engineer is like putting a loaded gun in the hands of a baby." The authors have seen misapplications of all prediction methods, including numerical models, and caution against anyone violating the requirement of a team knowledgeable in the sedimentation processes specific to the site and in the evaluation techniques applied.

The correct application of any evaluation technique—field, desktop, or model—requires a knowledge of the site, the tool, and its limitations, and a systematic, step-by-step approach as outlined in this manual. Further, interpreting and reporting the tool's results requires a thorough and explicit explanation of uncertainty bounds on the results. We endorse W. A. (Tony) Thomas's statement that "The purpose of modeling is insight, not answers."

We recommend use of the term *validation* for the process of using several observed data sets (bracketing the conditions to be tested) to demonstrate that a sedimentation model is useful for its intended purpose. The validation process must also produce well-defined uncertainty bounds to be reported in the interpretation of model results.

Finally, management of a navigation channel sedimentation study requires an organized approach just as systematic as the study's technical elements. Clear, written objectives, excellent client and team communications, and meticulous quality assurance/quality control are hallmarks of a well-managed study.

APPENDIX A
GLOSSARY

These terms are defined relative to solving sedimentation in navigation channels and their use in this manual.

ABS (Acoustic Backscatter Sensor): Measurement device of suspended sediment using the acoustic signal reflected from sediment particles

ADCP (Acoustic Doppler Current Profiler): Instrument that uses frequency shifts in returned acoustic signals to measure current velocities at various locations in flowing water

Advance Maintenance Dredging: Dredging beyond the authorized channel depth and/or width to prolong the period between required maintenance dredging procedures

Agitation Dredging: Removing sediment from the bottom of a channel and releasing it at the surface. The intention is that currents will sweep the agitated sediment away from the channel.

Anthropogenic: Originating in human activity

Antidune: Bedform that moves in a direction opposite to that of the flow by deposition on its upstream face and erosion of its downstream face

Avulsion (River): Sudden abandonment of a river of its existing channel and creation of a new channel

Base Test/Base Condition: Set of geometric configuration (bathymetry/hydrography and structures) and boundary and initial conditions (flows, sediments, etc.) with which other conditions are compared. Usually, the existing channel geometry and flow conditions

Base-to-Plan Test: Comparison of base test results to plan test results as relative changes, as in percent reduction in channel deposition in the plan test

Bed: Bottom of a waterway. Typically composed of sediment

Bedform: Irregularities in a sediment bed caused by interaction of flow and sediment

Bedload: Sediment transported by rolling, sliding, or jumping along the bed in response to flow and gravitational forces

Bed Material Load: Sediment in transport that is found in the bed at a location, either in suspension or as bed load

Brownian Motion: Random motions of tiny particles in a fluid caused by molecular collisions

Bypassing (Sediment): Process by which sediment is moved from one side of a transport obstacle to another; typically used to move sand from the updrift side of an inlet to the downdrift side

Calibration: Establishment of a one-to-one correspondence between the output of a measuring instrument's sensor, such as turbidity sensor, and the desired unit of measurement, such as NTU. Although calibration is widely used to describe the process of adjusting a model to reproduce observed data, we recommend against that usage for reasons given in the text.

Coldstart: Starting a numerical model with variables specified as simple initial values—such as a constant water surface elevation and suspended sediment concentration

Conceptual Site Model: Summary of various lines of evidence and conceptualization of dominant processes relevant for sedimentation at the project site

Coriolis: Effect that the rotating Earth has on objects or water moving large distances; from a fixed point on the Earth's surface, the object appears to curve to the right in the northern hemisphere. Numerical models of large water bodies must take this effect into account.

Desktop Model: Solution method that calculates quantities of interest (e.g., sediment deposition amount) from known or assumed quantities (e.g., flow discharge and channel geometry) using simple equations that can be solved by basic procedures such as running numbers on a calculator, a spreadsheet, or computer application

Digital Model: Digital representation of a waterway that includes inputs such as a grid, or mesh, that defines the physical boundaries of the waterway and initial elevations of the bed, banks, structures, and overbanks; some combination of water, sediment, and sometimes constituent inflows, outflows, and water levels at the water boundaries; bed and inflowing sediment characteristics; other driving forces information pertinent to the environment such as wind, waves, precipitation, and air temperature; parameters required by a numerical model computer program such as boundary roughness, turbulence closure coefficients, sediment characteristics, and so on

Draft: Depth of water occupied by a vessel from the water line to the lowest point of the vessel—usually the keel

Draghead: Device used on dredges to remove sediment from the bed of a waterbody. The draghead is positioned at the end of a dragarm through which water and sediment are suctioned.

Dune: Large bedform that moves in the direction of flow by erosion of its upstream face and deposition on its downstream face; also used for large subaerial sand features on land

Dynamic Draft: Depth of water occupied by a vessel from the water line to the lowest point of the vessel—usually the keel—when the vessel is in motion. Dynamic draft is greater than static draft.

Freshet: Flood of a river or stream, usually short-lived

GIS (Geographic Information System): Computer hardware and software that store, manage, analyze, and visualize spatial data

Geomorphologic: Relating to the physical features and processes of the Earth's surface

GUI (Graphical User Interface): Visual method of interacting with a computer program, usually involving computer screen windows, menus, and click-and-drag operations; in modeling, software system that facilitates the setup, development, testing and validation of numerical models through interactive computer screen displays

Grid (or Mesh): Network of points and areas or volumes used to provide a geometric framework for which variables can be input, calculated, and stored in a numerical model

Heisenberg Uncertainty Principle: In physics, Werner Heisenberg's uncertainty principle states that we cannot precisely know both the position and speed of an elemental particle. In modeling, we cannot know in advance precisely how well a model can be validated and how long it will take to achieve that level of validation.

Hotstart: Model computation that begins with initial conditions from a previous computation, such that instabilities and spin-up issues are minimized

Hydrostatic: Vertical pressure gradient is a function of gravity only. Vertical accelerations are not considered; applicable to most river, estuarine, and ocean flows, except near structures and abrupt changes in water level, such as surges and waves

Hysteresis: Lag between stage (water level) and discharge or between discharge and sediment load that produces a loop effect. For example, a given river stage may correspond to two different discharges—one for rising stage and another for falling stage.

Lacustrine: Related to lakes and reservoirs

Latin Hypercube Sampling: Method of selecting random values from a multidimensional space of possible values

Lightering: Procedure of removing part of a vessel's cargo to smaller vessels to reduce the larger vessel's draft, enabling it to enter shallower waters than its fully loaded draft

Lutocline: Location of a sharp density gradient caused by high sediment concentration

Mesh (or Grid): Ordered network of points and areas or volumes used to provide a geometric framework for which variables can be input, calculated, and stored in a numerical model

Model: Representation of a waterway that provides information about flow and sediment transport in the waterway

NTU (Nephelometric Turbidity UnitNTU): Measure of water turbidity

Numerical Model: Construct that combines a digital model of a waterway with a modeling program that solves equations of motion and transport. For example, "a numerical model of the Mississippi River" indicates a Digital Model of the Mississippi River combined with a Numerical Model Program.

Numerical Model Program: Ccomputer code that uses approximate methods to solve differential and algebraic equations describing water and sediment transport

Overdepth Dredging: Amount of dredging beyond specified channel depth for which the dredger is credited; used to ensure that full project depth is achieved

Physical Model: Scaled representation of a waterbody constructed to satisfy similitude laws

Physicochemical Processes: Combined physical and chemical forces acting on cohesive sediment grains

Plan Test/Plan Condition: Set of geometric configuration (bathymetry/hydrography and structures) and boundary and initial conditions (flows, sediments, etc.), which is to be evaluated for its effects. For example, does a proposed dike reduce model channel deposition compared with the model base test?

Porosity: Volume of voids divided by the total volume of sample

PDF (Probability Distribution Function): Variation of probability of a continuous random variable. For example, the PDF of a Normal (or Gaussian) random variable is a bell-shaped curve with a maximum at the mean value of the random variable.

Prototype: Original version; in modeling, the natural system that we model

Random Variable: Value that can be predicted only as a range of possible values with associated probabilities

Rheologic and Rheology: Related to relationships between forces and motion of a material. Typical sediment rheological measurement is the relationship between stress and strain

Ripple: Small bedform that moves in the direction of flow by erosion of its upstream face and deposition on its downstream face

Sand Wave: Bedform that moves in the direction of flow by erosion of its upstream face and deposition on its downstream face

Shoaling: Deposition of sediment in a navigation channel or change in shape or behavior of a wave as it propagates into water of decreasing depth

Specific Stage: River water surface elevation versus constant discharge over time; a constant relationship indicates equilibrium conditions, rising specific stage indicates net deposition in the reach, and falling specific stage indicates net erosion in the reach.

Squat: Amount that a floating vessel sinks relative to the water surface owing to its motion relative to the water's flow velocity

Static Draft: Depth of water occupied by a vessel from the water line to the lowest point of the vessel—usually the keel—when the vessel is at rest

Stochastic: Time series in which each of the values is a random variable. For example, suspended sediment concentration in a turbulent flow is stochastic—it consists of a mean value, a trend, and a randomly fluctuating component.

Suspended Load: Sediment transported in the water column, with infrequent bed contact

Taxonomy: Systematic classification

Thalweg: Line connecting the deepest points in each cross section of a waterway

Trim: Relative vertical positions of the bow and stern of a floating vessel. If they are equal, the vessel is "in trim."

Total Load: Quantity of all sediment in transport; sum of "Wash Load" and "Bed Material Load"

UKC (Underkeel Clearance): Distance between the lowest point on a vessel and channel bed

Validation: Process of ensuring that a model satisfactorily reproduces observed data (e.g., velocities, sediment deposition, and erosion); typically involves adjusting the model to fit several observed data sets representing a variety of flows and sedimentation processes. Some publications use terms such as *verification*, *confirmation*, and *circumstantiation* for this process.

Van der Waals Forces: Short-range electrical dipolar forces among uncharged particles

Verification: Process of ensuring that a numerical model program solves its equations correctly. Some publications use the term as a synonym for *validation*.

Wash Load: Sediment transported in suspension that is not found in the bed at that location

APPENDIX B
CASE STUDIES

These case studies illustrate a range of navigation channel types, sedimentation issues, and solutions along with the prediction methods used. Table B-1 lists the cases studies in the order they appear and shows classifications by geographic zone and channel draft.

Table B-1. Case Studies and Channel Categories.

Location	Inland	Estuarine	Coastal	Channel draft
Columbia River Mouth, Washington and Oregon		X	X	Deep-Draft
Lower Passaic River, New Jersey		X		Deep-Draft
Houston Ship Channel, Texas		X		Deep-Draft
New York/ New Jersey Harbor, New York and New Jersey				Deep-Draft
Norfolk Harbor, Virginia		X	X	Deep-Draft

Upper Mississippi River: Thebes to Ohio River Confluence, Illinois and Missouri	X			Shallow-Draft
Mississippi River Ship Channel, Louisiana	X	X		Deep-Draft
Cape Fear Entrance, North Carolina		X		Deep-Draft
South Shore Channels of Long Island, New York			X	Shallow-Draft
Tennessee–Tombigbee Waterway, Mississippi and Alabama	X			Shallow-Draft
Atchafalaya Bay, Louisiana			X	Deep-Draft
Dogtooth Bend Reach, Mississippi River, Illinois and Missouri	X			Shallow-Draft

MOUTH OF THE COLUMBIA RIVER

Setting

The deep-draft navigation channel at the Mouth of the Columbia River (MCR) between the states of Washington and Oregon in the Pacific Northwest, provides passage to international bulk cargo and container ships to six lower Columbia River Ports is shown in Figure B-1. The Columbia River Channel Improvement Project is 103 mi long from the Pacific Ocean to Portland, Oregon. The Federal Navigation Channel (FNC) was deepened between 1990 and 2010 to an authorized depth of 43 ft MLLW. Figure B-2 shows the Columbia River (MCR) Federal Channel A (north reach), which is 6 mi long, 2,000 ft wide and 55 ft deep, and Channel A (south reach), which is 6 mi long, 640 ft wide, and 48 ft deep. The MCR is protected by the Columbia River jetties. The North Jetty is 2.5 m long, the South Jetty is 6.6 mi long, and the Spur Jetty A is 0.3 mi long. These jetties are exposed to extreme waves of the Pacific Ocean and the sedimentation variability of the Columbia River and have experienced the need for significant repair over the last 20 years (USACE 2021).

Need for Sedimentation Prediction

Between 3 and 5 million yd^3 of dredging is required from shoals at the MCR federal navigation channels every year. These dredging requirements can be influenced by oceanic storm years that are larger than average, pushing sediment into the channels, and conversely, large Columbia River flows can reduce the need for dredging in the federal navigation channels by pushing sediment farther offshore, away from the channels. The

Figure B-1. Mouth of the Columbia River.
Source: USACE (2021).

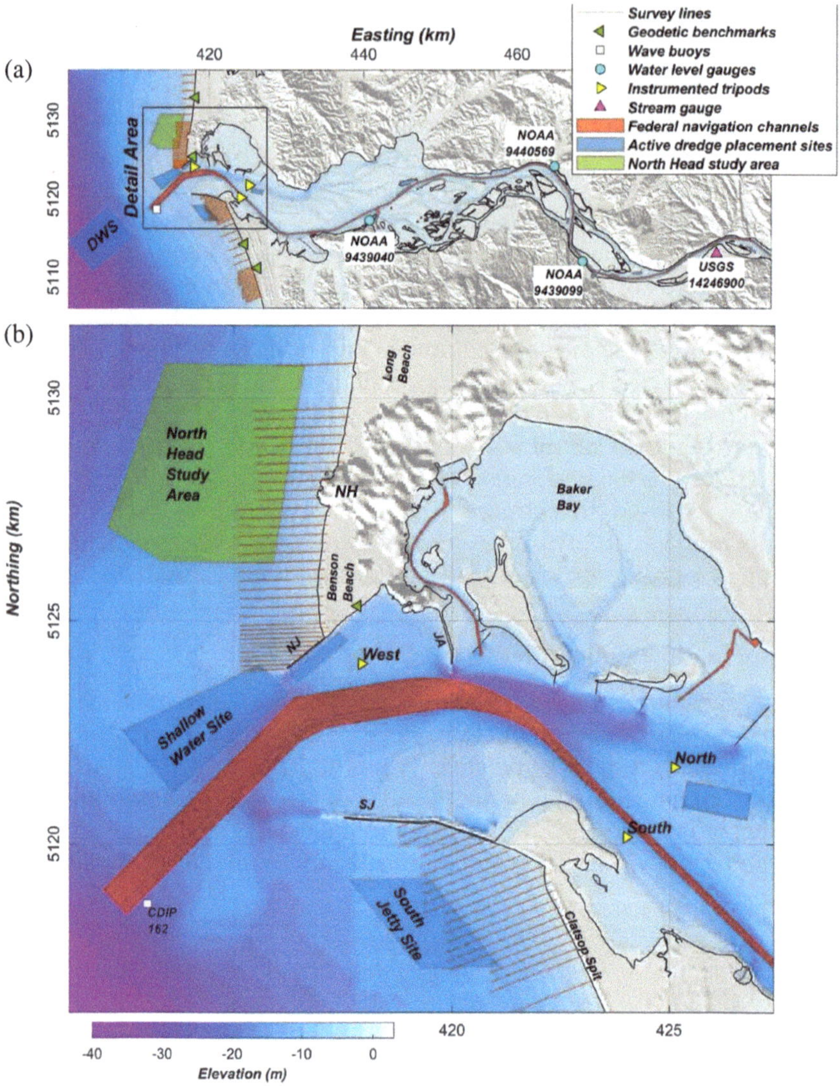

Figure B-2. Maps of Columbia River estuary (a) and mouth of Columbia River (b), showing locations of major engineering structures (JA, Jetty A; NJ, North Jetty; SJ, South Jetty). Blue shading shows active dredge-placement sites; green shading shows North Head study area; red shading shows maintained federal navigation channels. Also shown are locations of survey lines, geodetic benchmarks, wave buoys, water level gauges, instrumented tripods, and stream gauges. CDIP = Coastal Data Information Program; DWS = Deep Water Site; km = kilometer(s); m = meter(s); NH = North Head; NOAA = National Oceanic and Atmospheric Administration; USGS = US Geological Survey. Source: Stevens et al. (2020).

interaction of the jetty system with the navigation channel, the location and quantity of dredged material management and placement, and environment restriction and concerns all lead to the need for predictions of sediment movement, spatially and temporally within the MCR and FNC systems.

Sedimentation Prediction

The US Army Corps of Engineers Portland District has used a tiered approach for evaluation sediment transport and sedimentation prediction within the FNC (R. Moritz, personal communication, 2021). A large quantity of field data has been collected, including physical bathymetric-morphological conditions (shoaling locations and temporal accumulations) and environmental forcing data (oceanic and riverine hydrodynamics, flows, water levels and waves, etc.). On the basis of the data collected, the observed data and associated trends are used to assess FNC dredging needs, including shoaling locations and quantities. The primary circulation models used at MCR (ocean inlet) are Delft3D and ADCIRC. Within the LCR (Lower Columbia River upstream from ocean inlet), AdH Modeling is utilized.

For such a complex system such as the MCR, uncertainty in model outcomes is expected when using hydrodynamic and sedimentation numerical modeling tools, especially in the validation of the models. However, in certain cases, as needed to evaluate alternative FNC features (e.g., a pile dike field, variations in dredge sediment placement locations, and so on), a number of numerical models have been used for analysis. Circulation modeling has been performed using AdH, including the simple sediment transport indicator formulations using hydromodel results. Subsequent to AdH modeling, the Particle Tracking Model (Demirbilek et al. 2005) has been used, with either ADCIRC or AdH, defining sediment sources and sinks in base channel conditions and alternative channel conditions. In certain cases, Portland District has used the "full" sediment transport potential of the Delft3D model in baroclinic mode (bed load, suspended load components) for assessment of navigation channel feature alternatives and has used the AdH mobile bed sediment transport model (AdH-mobile bed).

The Portland District's goal with the tiered approach is to develop scientifically supportable analysis outcomes, so that informed decisions can be made about associated sediment transport trends predicted for various navigation channel temporal and spatial alternatives.

For the MCR and surrounding littoral system, Stevens et al. (2020) describe the use of both the observation of coastal change through the analyses of 6 years of repeated annual beach and nearshore morphology field surveys, and the numerical modeling of hydrodynamics, wave processes, and sedimentation using a validated coupled, Delft3D-SWAN to develop and evaluate the sediment budget in the study location and the understanding of sediment transport pathways. The hydrodynamic and sediment-transport model was run with a computational time step of 6 s. A visualization technique, the Sediment Pathway Interactive visualization Tool (SPIT), using the validated MCR hydrodynamic model, was used to show linkages and quantifications between alternative sediment source location and transport pathways. SPIT uses sediment-transport vector fields derived from a transport formula that was developed for noncohesive sediment transport, computed at each time step in Delft3D. Particle trajectories from 500 initial locations were modeled, under both two event-based time scales and an average annual conditions setup using 19 wave classes (inclusive of both summer and winter conditions), and high and low discharge scenarios. Figure B-3 shows interpreted sediment pathways from the SPIT results.

Elias and Gelfenbaum (2009) describe constructing a hydrodynamic and sediment transport model for the Mouth of the Columbia River using the Delft3D modeling system to examine and isolate the physical processes responsible for sediment transport and morphological change in this dynamic estuary entrance. The model is able to simulate the dominant

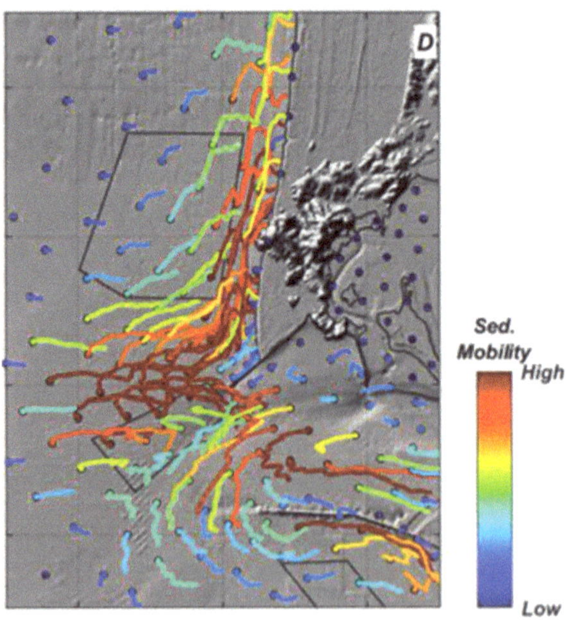

Figure B-3. Sediment Pathways around Columbia River Entrance as interpreted from SPIT results.
Source: Stevens et al. (2020).

features in the tidal flow, salinity, and wave fields observed in field measurements. Results show that a spatially and temporally varying balance of physical processes governs the transport of sand through the MCR. In the absence of salinity gradients, flows induce a large sediment export. Including salinity reduces the net sediment export at the Columbia River Mouth to near zero. Wave-driven transports are minor compared to the tides during typical summer conditions. In winter, during severe storm events, waves govern the transport patterns, increasing the transports by an order of magnitude.

REFERENCES

Demirbilek, Z., S. J. Smith, A. K. Zundel, R. D. Jones, et al. 2005. *Particle tracking model (PTM) in the SMS: III. Tutorial with examples*. Dredging Operations and Engineering Tech. Notes Collection (ERDC TN-DOER-D6). Vicksburg, MS: USACE Research and Development Center.

Elias, E., and G. Gelfenbaum. 2009. "Modeling processes controlling sediment transport at the mouth of the Columbia River." In *Proc., Coastal Dynamics 2009*. Paper No. 68. World Scientific, Singapore.

Stevens, A. W., E. Elias, S. Pearson, G. M. Kaminsky, et al. 2020. *Observations of coastal change and numerical modeling of sediment-transport pathways at the mouth of the Columbia River and its adjacent littoral cell*. US Geological Survey Open-File Rep. 2020-1045. Reston, VA: USGS.

USACE (US Army Corps of Engineers). 2021. *Fact sheet: Oregon coast: Mouth of the Columbia River*. Washington, DC: USACE.

LOWER PASSAIC RIVER, NEW JERSEY

Setting

The Lower Passaic River (LPR) is a highly urbanized and industrialized 28 km long estuary located in the state of New Jersey and is part of New York Harbor (Figure B-4). The dredging and deepening of the river for navigation purposes started in 1884, with nearly the entire length of the estuary dredged by 1924 (USACE 2010). The depth of the navigation channel has been increased progressively over time, until the last major dredging event in 1950 in the lower 11 km of the estuary, with depths

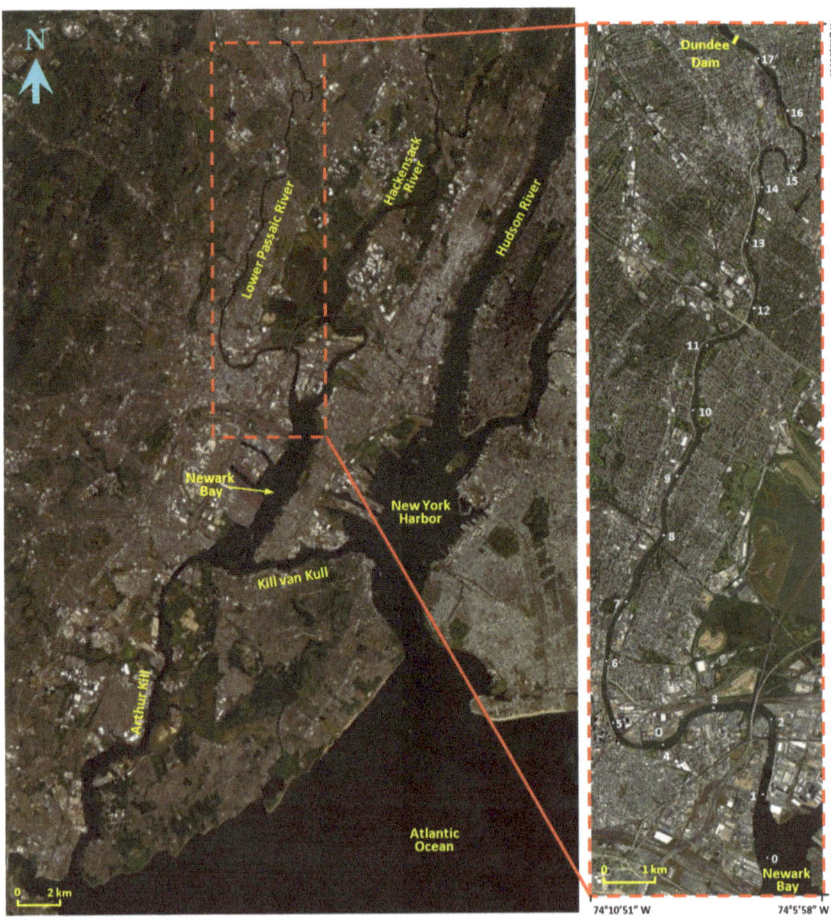

Figure B-4. Regional and local setting of the Lower Passaic River. Numbers along the longitudinal axis of the river in the inset panel indicate river miles.
Source: Mathew and Winterwerp (2022). Used with permission.

within the navigation channel several meters below the original preindustrial depths. Subsequently, and coincident with declining maritime use, the channel experienced infilling, with current bathymetry in some locations tending toward preindustrial conditions. Navigation is currently limited to the lower 3 km of the estuary.

Need for Sedimentation Prediction

Typical applications of various methods for estimating sedimentation rates involve relatively short-term or steady-state conditions. However, such methods may not be reliable in scenarios where the long-term progression of a project site is at question. This case study presents a computationally efficient numerical method suitable for performing long-term predictions of sedimentation rates and the overall morphodynamic progression of the system (Mathew and Winterwerp 2022). Potential scenarios where this methodology can be useful include the impact of sea-level rise, the impact of storms under a changing climate, impact of changes in sediment loadings or system forcings, and so on.

Sedimentation Prediction

A three-dimensional long-term time-variable numerical model was applied for estimating future sedimentation rates. Key processes represented in the model include the tide, salinity, temperature, river flow rate, resuspension, bedload transport, settling, deposition, consolidation, and flocculation (approximated in a parameterized manner).

The numerical model framework used the Estuarine, Coastal, and Ocean Model (ECOM) framework (Blumberg and Mellor 1987). The model was applied in morphodynamic mode (i.e., coupled hydrodynamics and sediment transport), with morphological change caused by erosion/deposition used to update bathymetry used in hydrodynamic calculations every time step. The computational time step ranged from 0.5 to 15 s, varying as a function of river flow rate (inversely related). The model framework was also extended to include morphological upscaling using the Morfac approach, which enables computationally efficient morphodynamic modeling. The morphological acceleration factor (applied to the computational time step and used to scale morphological response) ranged from 1 (during high river flow conditions) to 7 (during low river flow conditions).

Model performance was validated using an extensive short-term dataset of currents, water level, salinity, suspended sediment concentrations and fluxes, and bathymetric change (interannual periods). The validated morphodynamic model including the Morfac approach was applied to the 58-year period from 1950 to 2007 and model performance compared against the measured sedimentation over this period. The impact of navigation in the lower 3 km of the estuary was represented in a parameterized manner. The simulation spin-up duration was variable—for the short-term validation simulations, it ranged from a few months to a year, primarily to capture the effects of the antecedent river flow hydrograph on sediment inventory and suspended sediment dynamics during the following validation period. In contrast, for the long-term 58-year simulation, because the long-term morphological trend is insensitive to sediment spin-up duration, the sediment spin-up duration was limited to the hydrodynamic spin-up duration on the order of a few days. Overall, the model reproduces the spatial patterns of sedimentation during this period, thus indicating the applicability of the Morfac approach for performing long-term morphological simulations in fine sediment settings.

The morphodynamic model using the Morfac approach was applied in prognostic mode to assess, from a long-term (decadal scale) morphological perspective, the impact of ongoing navigation in the lower 3 km of the estuary and the impact of climate change (with sea-level rise and increased

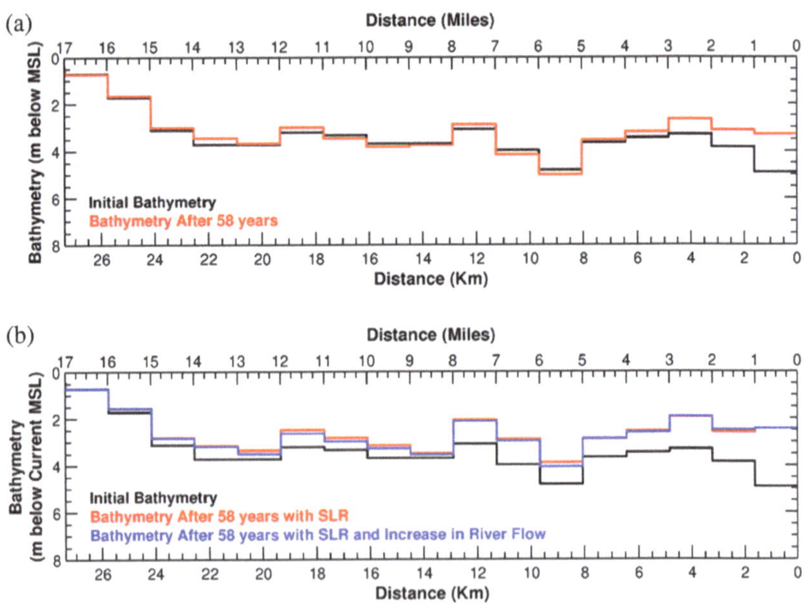

Figure B-5. Initial and model-calculated bathymetry after a 58-year prognostic simulation to assess (a) the impact of navigation in the lower 3 km of the estuary, and (b) the relative impact of sea-level rise and increasing river flow rate. Model results averaged laterally and longitudinally in 1 mi intervals.
Source: Mathew and Winterwerp (2022). Used with permission.

river flow assessed incrementally). Model results were evaluated in terms of morphological change, as well as common morphodynamic relationships of the local cross-section area with (1) location along the river, and (2) local tidal prism. The results show that ongoing navigation limits sedimentation in the lower 3 km of the estuary—in the absence of navigation, substantial sedimentation will occur, reaching an equilibrium morphology that is similar to that expected on the basis of morphodynamic relationships apparent in more landward reaches. Similarly, of the two impacts associated with global warming and climate change, sea-level rise rather than increased river flow rate has a greater impact on the morphological evolution of the estuary, with the overall impact of climate change being continued sedimentation. The results for both assessments are presented in Figure B-5.

REFERENCES

Blumberg, A. F., and G. L. Mellor. 1987. "A description of a three-dimensional coastal model." In *Three-dimensional coastal ocean models*. Coastal and Estuarine Sciences Series No. 4. edited by N. Heaps. Washington, DC: American Geophysical Union.

Mathew, R., and J. C. Winterwerp. 2022. "Morphodynamic modeling and morphological upscaling in a fine sediment system." *Adv. Water Resour.* 166 (August): 104224.

USACE (US Army Corps of Engineers). 2010. *Lower Passaic River commercial navigation analysis*. New York: USACE.

HOUSTON SHIP CHANNEL

Setting

The Houston Ship Channel (HSC) (Figure B-6) extends from the Bolivar Roads Channel just inland from the jetties at Galveston, to the port facilities in Houston, Texas. It is more than 55 mi long, and includes channel reaches ranging in authorized depth from 37.5 to 46 ft. The navigation channel is generally 530 ft wide. There are two 46.5 ft deep channels, which spur off of the main channel, to the container terminals at Bayport and Barbours Cut Terminal. The channel runs through Galveston Bay and within the San Jacinto River and the Buffalo Bayou. The Houston Ship Channel ranks second in the nation with respect to commercial tonnage and is estimated to provide $118 billion in annual economic benefit to the state of Texas.

The Houston Ship Channel is an estuary channel, with an approximate one-foot semidiurnal tidal range, surrounded by large-scale infrastructure, with multiple riverine inputs, and can be affected by hurricane storm surges and large precipitation events that can provide larger fresh-water inputs from Buffalo Bayou, and the Trinity and San Jacinto Rivers. The

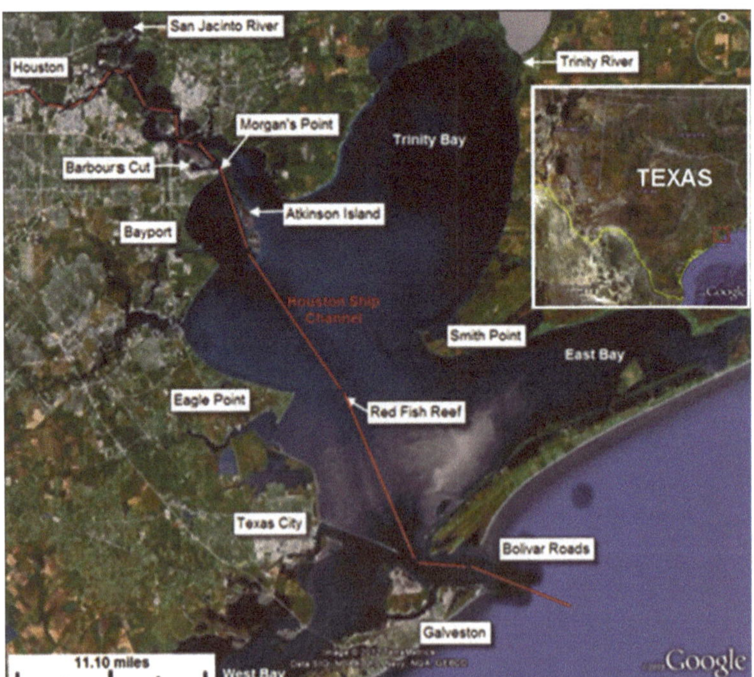

Figure B-6. Houston Ship Channel (HSC) area map.
Source: McAlpin et al. (2019b).

location is subject to wind-driven currents, especially in the shallow Galveston Bay. Geologic material varies from soft silt clays to very dense sand. The majority of the noncohesive materials in the channel are located at the south-eastern portion of the channel system, closer to the Gulf of Mexico.

Need for Sedimentation Prediction

The navigation channel sedimentation prediction was conducted to determine the ranges of volumetric shoaling changes that could occur with the modifications of channel deepening, channel widening, and channel bend easing, within six segments of the Houston Ship Channel for the Expansion Channel Improvement Project (ECIP). Predictions were conducted for the Existing Shoaling Rates, Estimated Shoaling Rates, Shoaling Estimate from a Numerical Model, and the Numerical Model Validation to the Corps Shoaling Analysis Tool (CSAT) Estimates (USACE 2019b).

Sedimentation Prediction

Existing Shoaling Rates were developed using quantities from historical and recent survey and dredging data (surveys available from 1999 to 2018, varied by channel segment per USACE 2019a), in some cases using nonpay dredging volumes, and in some cases not using the nonpay volumes (from the USACE Dredging Histories Database). A conservative contingency factor of 30% of volume estimates in certain "flare" locations of the channel system was used (USACE 2019a).

Estimated Shoaling Rates were developed by increasing the Existing Shoaling Rates by the same percentage as the footprint of the channel was to be increased (in sq ft). Where the channel footprint was not to be increased by the ECIP, the volume-of-cut method was used to estimate changes in shoaling rates (see "Desktop Approaches" in Chapter 5 in this manual for the strengths and limitations of these methods).

Shoaling Estimates originated from an adaptive hydraulics (AdH) numerical model of the estuary. The numerical model AdH was validated (McAlpin et al. 2019b) using the dredging records given in the USACE annual report by channel reach (USACE 2019b). The numerical model–computed shoaling results were scaled on the basis of historic dredge records for the existing condition (present without project—PWOP) and then calculated for three other conditions (present with project—PWP, future with project—FWP, and future without project—FWOP) in those channel locations that are not dominated by sand transport (McAlpin et al. 2019b). The 3D shallow water module of AdH version 4.6 was applied for all simulations. A 1-year spin-up period was used for water levels, salinity, and sedimentation. The time stepping was second order, and the maximum time step was 150 s. The results of the modeling show that the FWP alternatives increase channel sedimentation along most of the reaches of the HSC.

The sediment model was also validated with Corps Shoaling Analysis Tool (CSAT) estimates. (USACE 2019b). Repeated eHydro surveys of channel reaches were used to create input files for the CSAT database to provide estimates of shoaling rates on the basis of changes in bed elevation and in coordination with dredging events. The CSAT analysis for the HSC was conducted on the basis of data from 2011 to 2013. The CSAT data were compared to the Annual Report volume of dredging for the same time and then also used to scale the AdH model results of sedimentation for the alternative project conditions listed previously (PWP, PWOP, FWP, FWOP). This validation method provided a much larger range of possible sedimentation rates for the HSC than the Annual Report values for the modeled period. The two methods (AdH and AdH scaled with CSAT analysis) present a possible range of sedimentation under various flow conditions along sections of the HSC.

REFERENCES

McAlpin, J., C. Ross, and J. McKnight. 2019a. *Houston ship channel and vicinity three dimensional adaptive hydraulics (AdH) numerical model calibration/validation*. HSC-ECIP Engineering Appendix C 17-2, Rep. ERDC/CHL TR-19-10. Vicksburg, MS: USACE, Coastal and Hydraulics Laboratory, Research and Development Center.

McAlpin, J., J. McKnight, and C. Ross. 2019b. *Houston ship channel expansion channel improvement project (ECIP) numerical modeling report*. ERDC/CHL TR-19-12. Vicksburg, MS: USACE, Coastal and Hydraulics Laboratory, Research and Development Center.

USACE (US Army Corps of Engineers). 2019a. *Houston ship channel expansion channel improvement project, Harris, Chambers, and Galveston Counties, Texas*. Engineering Appendix C, Galveston District, Southwestern Division. Washington, DC: USACE.

USACE. 2019b. *Houston ship channel expansion channel improvement project, Harris, Chambers, and Galveston Counties, Texas*. Engineering Appendix C, Attachment 8, Corps Shoaling Analysis Tool (CSAT) Report, Galveston District, Southwestern Division. Washington, DC: USACE.

NEW YORK/NEW JERSEY HARBOR

Setting

The New York/New Jersey Harbor (NYNJH) is an estuarine deep-draft navigation channel system, from the entrance at the Verrazano-Narrows, through Lower New York Bay, with channels in the Arthur Kill, within Newark Bay, adjacent to port facilities, and north up the Hudson River.

Presently, the main navigation channel has an authorized depth –50 ft MLW, with side channels having depths of –40 ft or –45 ft MLW. The channel system has more than 100 mi of shipping channels, ports and anchorages, including 31 mi of deep-draft channels. The Port of New York and New Jersey moves cargo with a value of more than $50 billion every year. Figure B-7 shows the modeling representation of the various channels and surrounding bathymetry, with the vertical layer scheme used in the NYNJH digital model mesh.

The Harbor flow regime includes an approximate 5 ft semidiurnal tidal range, multiple significant riverine inputs, and flow from sewage treatment

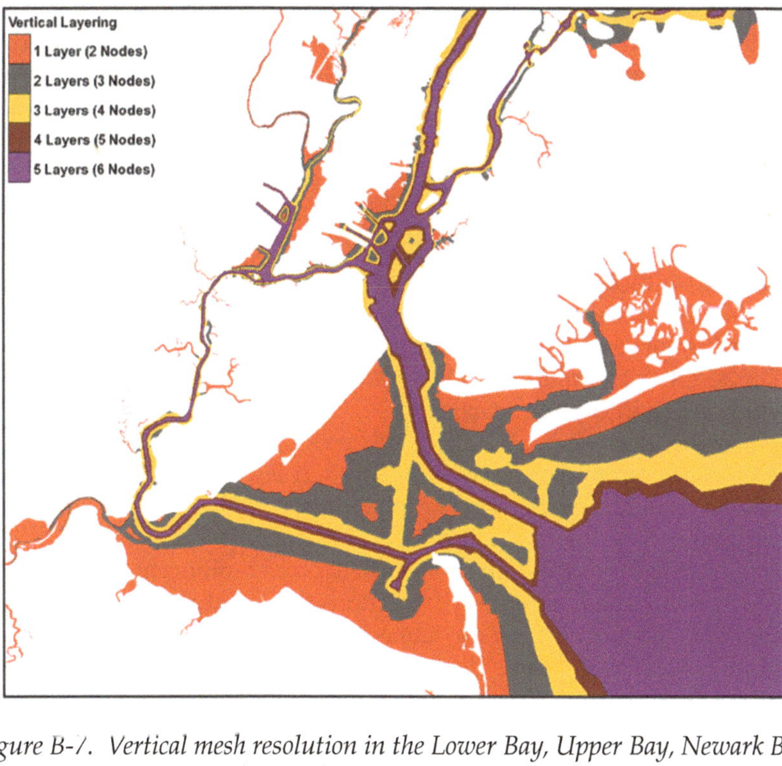

Figure B-7. Vertical mesh resolution in the Lower Bay, Upper Bay, Newark Bay, and adjacent NY–NJ Harbor study areas.
Source: McAlpin et al. (2020).

plants. The system is affected by waves, currents, and high water levels from tropical and extratropical storms. Wind-generated waves and deep-draft ship transit and associated bow waves have an influence on adjacent bay shorelines and sediment movement within the channel limits. The area has cohesive and noncohesive sediments, and organic sediments with cohesive properties. Dredging operations within the channel system can also influence sedimentation patterns.

Need for Sedimentation Prediction

A study of channel sedimentation prediction was completed in 2020 (McAlpin et al. 2020). The purpose of the numerical modeling study was to determine sedimentation rates in various reaches of the –50 ft MLW with-project deepened harbor channels (compared to the preproject condition of –45 ft MLW in the main channels), owing to influences of the following: sediment sources, channel deepening, boat wakes, sediment movement caused by coastal storms and riverine high flows. The numerical model results were analyzed to provide insight into which locations and channel reaches will experience increased/decreased deposition and quantify the amount of increase/decrease for a given channel reach, on the basis of significant natural sedimentation variability in the project area and impact owing to channel deepening.

Sedimentation Prediction

Five years (1985, 1995, 1996, 2011, and 2012) were simulated for both the with- and without-project conditions to determine the impact of the channel deepening on the dredging requirements for a wide range of meteorological conditions, including storm events.

The adaptive hydraulics (AdH) model was used with SWAN for waves generation, incorporating an existing validated ADCIRC modeling grid for the region, and the CSTORM-MS modeling system. A 1-month hydrodynamic spin-up time was employed. A 1-year sedimentation model simulation (2012 forcings) was completed to spin up the bed composition without allowing the bed elevations to change. Internal time-step size adaption allows for time-step changes to capture rapidly changing physics during run time. Sedimentation predictions were evaluated using SEDLIB, a sediment transport library developed at ERDC (Brown 2012a, b).

Model validation considered the following: Sources of model uncertainty and consequences, two-dimensional (2DH) tidal harmonic comparisons in the contributing rivers, three-dimensional (3D) hydrodynamic comparisons, wave model validation, quantitative and qualitative 3D salinity transport comparisons, quantitative and qualitative

3D sediment transport comparisons, dredge volume comparisons, sediment core comparisons, and qualitative sediment flux comparisons. The variations of each of these sources of input forces or values were varied to determine changes in sedimentation quantities and patterns from preproject to with-project conditions: sewage flow, diffusion specification, friction specification, wind wave sensitivity, riverine inflow sensitivity, sea-level rise sensitivity, and initial bed specification.

The model results indicated a relatively minor increase in the total sedimentation and resulting dredge volumes for the NYNJH under the with-project –50 ft MLW condition, with the increase being insignificant in comparison to the natural variability in sedimentation/dredge volumes across years owing to environmental, hydraulic, and meteorologic forces.

REFERENCES

Brown, G. L. 2012a. "A quasi-3D suspended sediment model using a set of correction factors applied to a depth averaged advection diffusion equation." In *Proc., IIHR 3rd Int. Shallow Flows Symp.* Iowa City, IA: University of Iowa.

Brown, G. L. 2012b. "Modification of the bed sediment equations of Spasojevic and Holly (1993) to account for variable porosity, variable grain specific gravity, and nonerodable boundaries." In *Proc., IIHR 3rd Int. Shallow Flows Symp.* Iowa City, IA: University of Iowa.

McAlpin, T. O., J. V. Letter Jr., M. Bryant, A. G. Emiren, et al. 2020. *New York/New Jersey Harbor sedimentation study numerical modeling of hydrodynamics and sediment transport*. ERDC/CHL TR-20-15. Vicksburg, MS: USACE Coastal and Hydraulics Laboratory, Research and Development Center.

NORFOLK HARBOR, VIRGINIA

Setting

The navigation infrastructure within Norfolk Harbor (NH) comprises a deep-draft navigation channel network connecting ports in the cities of Norfolk, Newport News, Portsmouth, Chesapeake, and Hampton in southeastern Virginia to the Atlantic Ocean through Chesapeake Bay (Figure B-8). The physical setting of the channel network comprises a predominantly coastal regime (i.e., tides and waves) in the outer reaches, estuarine in the more landward reaches, and tidal in the intermediate reaches. As part of an investigation to address operational limitations in commercial and defense vessel activities, the USACE conducted a feasibility study assessing the widening and deepening of a 59 mi portion of the channel network in Norfolk Harbor.

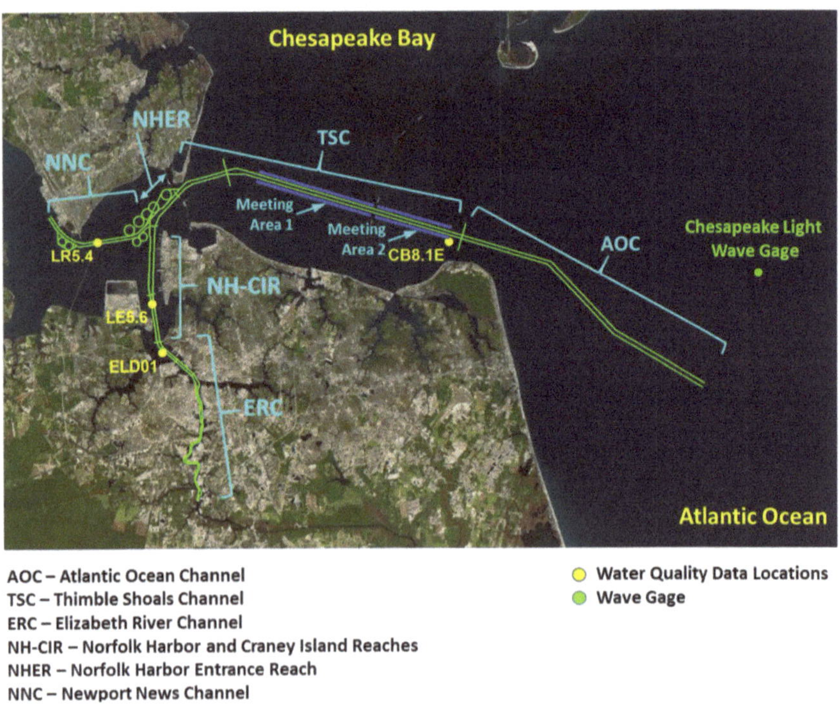

Figure B-8. Navigation channel layout of Norfolk Harbor and Elizabeth River and locations of various data sources used in sedimentation analysis.
Source: Adapted from Moffatt & Nichol (2018).

Need for Sedimentation Prediction

Existing channel depths (proposed depths in parentheses) are 35 to 40 ft (37 to 47 ft) in the ERC reach, 50 ft (58 ft) in the NH-CIR, NHER, NNC, and TSC reaches (Figure B-8), and 52 ft (61 ft) in the AOC reach. Relative to natural channel depths (i.e., depths in the portion of the cross section outside the channel) that range from about 15 ft in the ERC reach to 44 ft in the AOC reach, the proposed channel depths represent a significant deepening in some reaches. For instance, the channel bottom in the AOC reach is currently only 8 ft deeper than the adjacent seafloor outside the channel but is projected to be 17 ft deeper than the adjacent seafloor in the future. The doubling of relative channel depth represents a significant increase in sediment trapping efficiency. Therefore, the feasibility study included an assessment of projected sedimentation rates (and effectively, the projected maintenance dredging rates) following the channel modifications (Moffatt & Nichol 2018).

Sedimentation Prediction

Several desktop analytical methods were used to estimate future sedimentation rates in the various reaches, depending on the key hydrodynamic processes responsible for sediment transport. In the ERC and NH-CIR reaches, tidal infilling and density (salinity)-driven circulation dominate, whereas in the NHER and NNC reaches, tidal currents dominate, and in the TSC and AOC reaches, tidal currents and waves dominate. Sediment substrate varies accordingly as well, with the AOC reach comprised of predominantly sandy sediments, TSC reach comprised of a mix of fine sediments and sands, and the remaining more landward reaches dominated by fine sediments.

Sedimentation in the ERC and NH-CIR reaches was estimated using the harbor basin sedimentation approach of Eysink and Vermaas (1983) and Eysink (1989). Sedimentation in the NHER, NNC, TSC, and AOC reaches was estimated using the channel sedimentation approach of Van Rijn (2013). In addition, sedimentation in the TSC reach was also estimated using the empirical Volume-of-Cut approach (Van Rijn 2013) as a confirmatory check on the channel sedimentation approach.

Both the harbor sedimentation and the channel sedimentation analytical approaches were implemented using spreadsheet models. Data required for applying these models include information on channel geometry, salinity, suspended sediment concentrations, tidal currents, water depths, wave climate, and current maintenance dredging rates. Analytical models were developed using the current channel geometry and validated to the current maintenance dredging rates and applied to estimate future sedimentation rates using the proposed channel geometry.

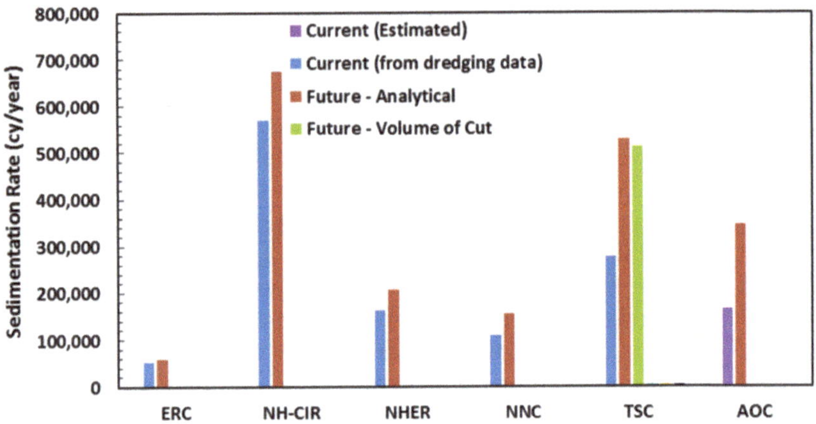

Figure B-9. Current and estimated future sedimentation (maintenance dredging) rates using analytical and empirical approaches.
Source: Adapted from Moffatt & Nichol (2018).

Figure B-9 shows a comparison of the current and estimated future sedimentation rates. Overall increases in sedimentation rates are relatively modest in the landward reaches, ranging from 11% in the ERC reach to 42% in NNC reach. However, the TSC and AOC reaches are estimated to experience a near-doubling of sedimentation rates, primarily because of the doubling of the channel depth (relative to the adjacent seafloor) compared to current conditions.

REFERENCES

Eysink, W. D. 1989. *Sedimentation in harbour basins: Small density differences may cause serious effects*. Publication No. 417. Delft, The Netherlands: Delft Hydraulics.

Eysink, W. D., and H. Vermaas. 1983. *Computational methods to estimate sedimentation in dredged channels and harbour basins*. Publication No. 307. Delft, The Netherlands: Delft Hydraulics.

Moffatt & Nichol. 2018. *Desktop assessment of future sedimentation rates*. Engineering Appendix D: Sedimentation Study to Engineering Appendix A in report titled General Reevaluation Report and Environmental Assessment, Norfolk Harbor Navigation Improvements. Norfolk, VA: USACE.

Van Rijn, L. C. 2013. "Basics of channel deposition/siltation." www.leovanrijn-sediment.com. Accessed 1 November 2022.

UPPER MISSISSIPPI RIVER: THEBES, ILLINOIS TO OHIO RIVER CONFLUENCE

Setting

The Mississippi River between Thebes, Illinois, and the confluence with the Ohio River (UMR) has experienced three significant floods in the last decade. The 2011 flood produced a historic peak at Cairo, Illinois, gauge and for the first time since the flood of 1937, the Birds Point–New Madrid Floodway was activated. The 2015 flood on the Upper Mississippi (above the confluence with the Ohio River) produced historically high discharges of 29,700 cm (1,050,000 cfs) at Thebes, Illinois, and resulted in a breach of the non-federal Len Small levee near Miller City, Illinois, on January 2, 2016 (Figure B-10). Finally, the 2017 flood on the Upper Mississippi River produced discharges at Thebes, Illinois, of 26,300 cm (928,000 cfs) and flows that continued to drain through the still-open crevasse of the Len Small levee.

Need for Sedimentation Prediction

The failure of the Len Small levee resulted in significant erosion to farmland across the Dogtooth Island Peninsula (DIP) and represented the first stages of a meander cutoff. A numerical modeling study was

Figure B-10. Aerial photo of the Len Small breach, looking east-southeast through the breach, January 2, 2016.
Source: Courtesy of Dynamic Solutions, LLC. Used with permission.

conducted to investigate hydraulics and sediment transport in the river channel and across the DIP to provide information to estimate the increased risks of future disruptions to navigation.

Sedimentation Predictions

The USACE's adaptive hydraulics (AdH) numerical modeling program was selected for use on this study, described by Brauer et al. (2021). The AdH software was specifically developed to simulate the complex hydrodynamics and resulting sediment transport associated with levee and dam failures.

An important AdH feature is its ability to automatically adapt the mesh in areas where additional spatial resolution is needed to properly resolve the hydrodynamics. AdH can also expand elements in previously refined mesh areas where higher resolution is no longer needed. AdH is capable of simulating overland flow and two (2D) or 3D shallow water problems. For this application, the 2D shallow water (SW) module was utilized. The sediment transport simulation capability within AdH is facilitated through linkage to the USACE sediment transport library (SEDLIB).

A model mesh including 63,464 computational nodes and 126,274 elements (without adaption) was developed to represent the Upper Mississippi River, from Thebes, Illinois, to Birds Point, Missouri (Figure B-11). Typically, there were 15 to 22 nodes across the channel with at least 12 nodes across the low-flow channel. The channel elements in the Mississippi River were generally 30 to 45 m wide.

The hydrodynamic model was developed and validated to data from periods in 2011, 2015 to 2016, and 2017 floods. The 2011 simulation period was from April 15, 2011, to May 31, 2011. The 2015 to 2016 simulation period was from December 19, 2015, to January 16, 2016. The 2017 simulation period was from April 23, 2017 to May 20, 2017.

The DIP sediment transport model results were validated to observed areal averaged elevation changes and resulting changes in sediment volume for the 2015 and 2017 floods.

The AdH model was used to simulate a pre-2019 baseline alternative condition and three structural alternatives along a highway (Watson Road) designed to simulate the effects of flow and sediment transport across the DIP. A second post-2019 baseline alternative condition was also developed and compared to a 2020 current condition alternative and a tree-screen alternative.

An examination of the three Watson Road structural alternatives showed reduced flows over the DIP, but the reduction in flows had little to no effect on velocity, shear stress, or bed displacement once the structures were overtopped. For the tree-screen alternative, which simulated trees over much of the DIP flow pathways, flow through the Len Small breach

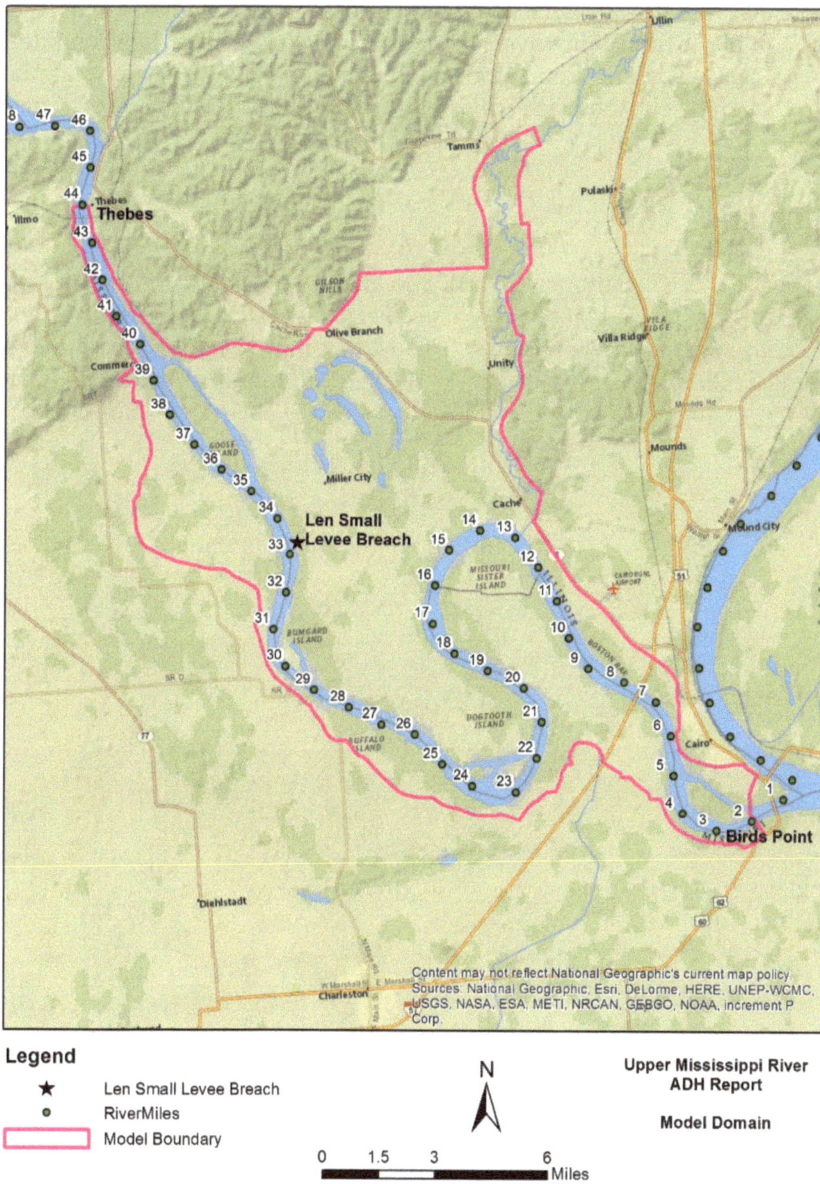

Figure B-11. Upper Mississippi River AdH Model domain.
Source: Courtesy of Dynamic Solutions, LLC. Used with permission.

decreased by 10% to 20% and shear stresses decreased by 3 to 10 Pa through major flow pathways. The tree-screen alternative, when compared to the 2020 Current Conditions in the major flow pathways, almost eliminated the erosion, with only small areas indicating erosion of 0.5 m or less.

REFERENCE

Brauer, E., S. C. Sanborn, W. H. McAnally, and C. M. Wallen. 2021. *Mississippi River AdH model modification and valuation, Thebes, IL to Birds Point, MO Reach*. Washington, DC: Mississippi River Geomorphology and Potomology Program, USACE, Lower Mississippi Valley Division.

MISSISSIPPI RIVER SHIP CHANNEL

Setting

The Mississippi River Ship Channel (MRSC), Gulf to Baton Rouge, Louisiana, is a federal deep-draft navigation project beginning at Baton Rouge and terminating at the Gulf of Mexico, more than 250 mi downriver (Figure B-12). The primary purpose of the project is to provide deep-draft access to Louisiana ports along the Mississippi River; four of these ports are in the top thirteen in the United States (USACE 2018).

Need for Sedimentation Predictions

Presently, 12 crossings upriver of New Orleans are routinely dredged, and three are occasionally dredged, to a depth of 45 ft below Low Water Reference Plane (LWRP), plus 3 ft of advance maintenance and 2 ft of overdredging (USACE 2018). Dredging normally begins on the recession side of the flood hydrograph after shoaling has begun. Two crossings have dike fields to reduce maintenance dredging needs. During fiscal years 2007

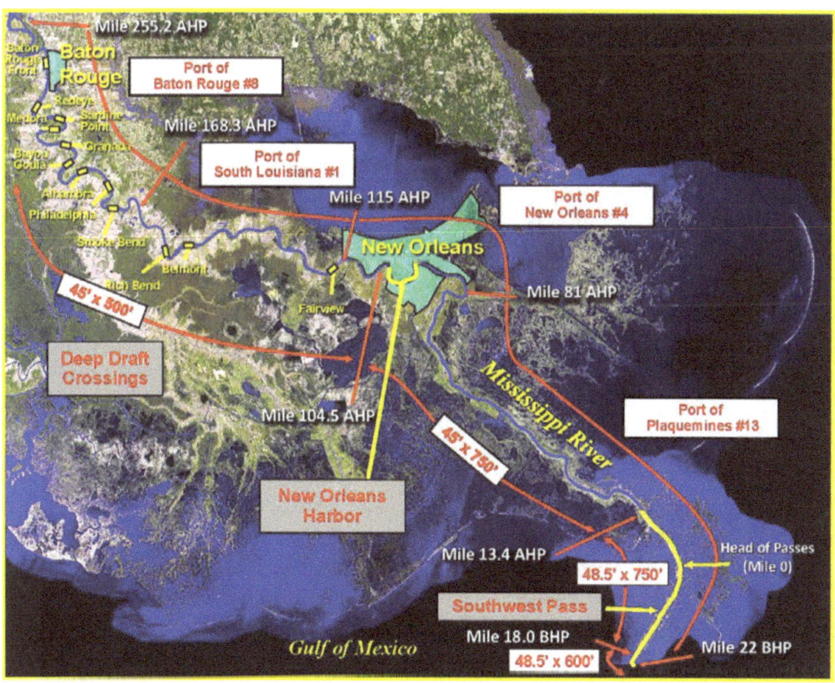

Figure B-12. Mississippi River ship channel project current dimensions. Source: USACE (2018).

to 2016, an average 22.4 million yd^3 (mcy) has been dredged annually upriver of New Orleans (USACE 2018).

Near Venice, in the Head of Passes (HOP) area, and in Southwest Pass (SWP), portions of the ship channel are dredged to a depth of 48.5 ft below Mean Lower Low Water (MLLW), plus 6 ft of advance maintenance and 2 ft of overdredging (USACE 2018). Dredging is performed during the flood hydrograph when shoaling occurs. Fiscal year 2007 to 2016 data show that an average annual 16.9 mcy of material have been dredged from HOP and SWP, including the SWP Bar Channel (USACE 2018).

After completion of a feasibility study and environmental impact statement by the US Army Corps of Engineers (USACE), the 1985 Supplemental Appropriation authorized a 55 ft draft navigation channel. The project was partially implemented to the current ship channel dimensions. A General Reevaluation Report (GRR) and Supplemental Environmental Impact Statement were completed by USACE in 2018, looking at whether additional deepening is in the Federal interest.

Sediment supply, hydrograph shape, channel bank elevations, and distributary/diversion flows are some of the major influences on the location and timing of shoaling in the Mississippi River. A saltwater wedge is present in SWP year-round and can move upriver during low flow; the location of the saltwater interface also affects the location of shoaling. In-channel sediment storage can occur; this sediment may be transported downriver during the next flood hydrograph.

Sedimentation Predictions

USACE feasibility study economic and environmental analyses used historical dredging records and in-situ volume computations to estimate/predict the change in maintenance dredging for the crossings and HOP with an increase in project depth from 40 ft to 45, 50, or 55 ft. For SWP, the increase in maintenance dredging was based on historical dredging records and the net effects of the increase in cross-sectional area and the increase in SWP flow as a result of construction of training works such as foreshore and spur dikes (USACE 1981). Some of these training works were not implemented (USACE 2018).

During the design phase in the late 1980s, USACE used several numerical models to assess the effect of potential project features on shoaling or maintenance dredging in HOP and SWP, including increasing advance maintenance depth, sediment trap construction, bank protection works restoration, and dredging and flow diversion strategies at Cubits Gap (Copeland and Thomas 1992; Richards and Trawle 1987, 1988a, b; Lin et al. 1990).

For Redeye Crossing, an existing one-dimensional numerical model was selected as most expedient and modified to consider the concept of dikes and their effect on channel shoaling and dredging (Copeland and Thomas 1992).

For the GRR, USACE modified existing 1D and 2D numerical models to enhance their shoaling prediction abilities; these models provided information to determine changes to maintenance dredging with a deeper channel (Copeland et al. 2020, Sharp et al. 2013, USACE 2018).

In general, 1D models were used for long simulation periods because of their computational efficiency and 2D models were used to provide details unavailable in a 1D model. Model program selection depended on site- and project-specific requirements for model capabilities, model programs available at the time, and to some extent, on modelers' familiarity with individual modeling programs. As noted in the *Tool Selection* section of Chapter 6, sometimes the best model program is the one that the modeler is most familiar with, and some of the selections may have been based on that factor.

Pertinent information on these models is presented in Table B-2.

Efforts were made to identify model limitations and uncertainties and, in some cases, adjust for them. The numerical models listed in Table B-2 are all either 1D or 2DH and therefore not capable of addressing the mechanics of shoaling in a saltwater wedge environment. In the 1D models, deposition coefficients for silt and clay were adjusted so that calculated dredging quantities duplicated SWP dredging records; this may have captured the outcome of the existing saltwater wedge environment for the tested hydraulic conditions; however, without modeling the mechanics, the effect of channel deepening, sea-level rise, and river flow changes on the location of the saltwater wedge and its subsequent effect on shoaling cannot be determined from the model results. A 2D vertical model or 3D model is necessary to evaluate those conditions.

To compensate for Tabs-2 software's one sediment size class, different Manning's *n* values for bed shear stress were specified for the rising and falling limbs of the hydrograph. The AdH model did not consider silts and clays in the modeling; the results were adjusted to better represent the portion of dredged material comprised of silts and clays. For the crossings, the AdH model predicted cumulative volumes of dredged material well but did not consistently predict the distribution among the individual crossings. Nor did the AdH modeling depict dredging as it is conducted. A sensitivity index was calculated equal to the volume dredged with continuous dredging divided by the volume dredged for one dredging event per year; this sensitivity index represented dredging uncertainty in the modeling, which was addressed in how the results were presented.

Model results were presented in ways that could be used with observational data to predict shoaling or dredging. For Tabs-1 simulations, model results were presented in terms of dredging volume changes and percentages. Tabs-2 modeling results were presented in terms of percent

Table B-2. Mississippi River Ship Channel Numerical Modeling.

Model	Domain	Model output area	Sediment transport function	Sediment size/classes	Dredging parameters	Simulation period
Tabs-1[a]	Tarbert Landing to SWP jetties	HOP and SWP	Laursen, modified by Madden	Eight classes, medium sand through clay	Recession side of hydrograph when flow below 700,000 cfs, material removed from model	32 years, from 1975 through 2006
Tabs-1	Tarbert Landing to SWP jetties	Redeye Crossing	Laursen, modified by Madden	Eight classes, medium sand through clay	Recession side of hydrograph, dredging capacity and initiation time specified, material returned to water column	15 years, from 1975 through 1989
Tabs-2	Venice to GOM	HOP and SWP	Ackers-White	0.15 mm	Dredging not performed	Steady state
Tabs-2	Venice to GOM, channels only	HOP and SWP	Ackers-White	0.15 mm	Dredging not performed	87 days

(Continued)

Table B-2. (*Continued*).

Model	Domain	Model output area	Sediment transport function	Sediment size/classes	Dredging parameters	Simulation period
HEC-6T	Tarbert Landing to SWP jetties	HOP and SWP	Toffaleti-Meyer–Peter Muller For silt and clay, Krone and Partheniades adapted by Ariathurai and Krone	15 classes, very coarse gravel to clay	Deposition exceeds a trigger elevation. For the crossings, also only on recession side of hydrograph and flow below 600,000 cfs. Material returned to water column (crossings) or removed from model	50 years, from 2020 to 2069
AdH	Old River to GOM	Crossings upriver of New Orleans	van Rijn, modified by Kleinhans and van Rijn, and Wright and Parker	Seven classes, fine gravel to very fine sand	Dredging once a year, or continuously. Material returned to water column	3-year periods representing years 2025 and 2075

[a]Tabs-1 was a 1D predecessor to HEC-6T.

change in shoaling rate per month, total shoaling, average shoaling per day, and maximum shoaling per day. A shock shoaling index was also developed by dividing the maximum daily shoaling by the average daily shoaling; this index was used to determine whether the dredge plant could keep up with the peak shoaling rates. The HEC-6T model results were presented in terms of average annual dredging quantities over a 50-year simulation and a dredging index, the ratio of the computed dredging quantity at a specific location and a base condition, which could be applied to dredged material data. For AdH modeling, the model results for each plan were also presented in terms of a dredging index: the volume of dredged material for the plan divided by the volume of dredged material for the base for the same boundary conditions. The individual crossing dredging indexes were accumulated into a cumulative dredging index, which could be applied to dredged material data. In the lower river, the dredging index for sand was combined with an assumed index of one for the silts and clays by computing a weighted average of the dredging index for each reach weighted by the fraction of sand deposited by reach approximated from the HEC-6T results.

For future conditions, the Tabs-2 modeling included a scenario where SWP overbanks would subside 0.5 ft. For the AdH model, following USACE guidance documented in USACE (2019), three different eustatic sea-level rise values were added to a spatially and temporally constant downstream stage boundary for year 2075 simulations. For year 2025 simulations, because the range of values was small, a single value of eustatic sea-level rise was added. The effects of subsidence and morphological changes between years 2025 and 2075 were neglected. To address uncertainty in subsidence, one run for year 2075 was made with one eustatic sea-level rise added to the downstream boundary and full subsidence included in the bed elevations. For the HEC-6T modeling, simulations for each channel depth were conducted with no eustatic sea-level rise and for two rates of eustatic sea-level rise. The model included an estimate of spatially varying subsidence rates that were automatically applied to model geometry during the simulation. In both AdH and HEC-6T modeling, dredging templates were kept constant and not changed because of increasing stages.

REFERENCES

Copeland, R. R., L. Lombard, and R. A. Gaines. 2020. *Numerical sedimentation investigation Mississippi River Cairo to Pilots Station.* Mississippi River Geomorphology and Potamology Program, Rep. No. 30. Washington DC: US Corps of Engineers.

Copeland, R. R., and W. A. Thomas. 1992. *Lower Mississippi River Tarbert Landing to east jetty sedimentation study*. Tech. Rep/HL-92-6. Vicksburg, MS: Dept. of the Army, Waterways Experiment Station, USACE.

Lin, H. J., W. D. Martin, and D. R. Richards. 1990. *Dredging alternatives study Cubits Gap, Lower Mississippi River*. Rep. 2, Tabs-2 Numerical Model Investigation, Vols. 1 and 2, Tech. Rep. HL-90-20. Vicksburg, MS: Dept. of the Army, Waterways Experiment Station, USACE.

Richards, D. R., and M. J. Trawle. 1987. *A numerical model analysis of Mississippi River passes navigation channel improvements*. Rep. 1: 55-Foot Channel Tests, MP-HL-87-2. Vicksburg, MS: Dept. of the Army, Waterways Experiment Station, USACE.

Richards, D. R., and M. J. Trawle. 1988a. *A numerical model analysis of Mississippi River passes navigation channel improvements*. Rep. 2: 45-ft Channel Tests and Flow Diversion Schemes, MP-HL-87-2. Vicksburg, MS: Dept. of the Army, Waterways Experiment Station, USACE.

Richards, D. R., and M. J. Trawle. 1988b. *A numerical model analysis of Mississippi River Passes navigation channel improvements*. Rep. 3, Bank Breaching without Supplement 2, MP-HL-87-2. , Vicksburg, MS: Dept. of the Army, Waterways Experiment Station, USACE.

Sharp, J., C. Little, G. Brown, T. Pratt, et al. 2013. *West Bay sediment diversion effects*. ERDC/CHL TR-13-15. Vicksburg, MS: USACE, Engineering Research and Development Center.

USACE (US Army Corps of Engineers). 1981. *Deep-draft access to the ports of New Orleans and Baton Rouge, Louisiana*. Feasibility Study, Vol. 3, Appendix C, Engineering Investigations. New Orleans: USACE.

USACE. 2018. *Mississippi River Ship Channel, Gulf to Baton Rouge, LA*. Integrated General Reevaluation Report and Supplement III to the Final Environmental Impact Statement, Mississippi River Ship Channel, Gulf to Baton Rouge, Louisiana Project. New Orleans: USACE, Mississippi Valley Division.

USACE. 2019. *Procedures to evaluate sea level change: Impacts, responses, and adaptations, pamphlet*. EP 1110-2-1. Washington, DC: Dept. of the Army, USACE.

CAPE FEAR RIVER, NORTH CAROLINA

Setting

The Cape Fear River (CFR) is located in southeastern North Carolina. The federally authorized Cape Fear River Navigation Channel connects the Atlantic Ocean to the Port of Wilmington, NC. The navigation channel is 38 mi long, has an authorized channel depth, plus an allowable overdepth ranging from 36 to 46 ft MLLW. It is a deepwater port and is a major economic contributor to the southern North Carolina region. Figure B-13 shows the Cape Fear River ocean entrance channel and the adjacent barrier island shorelines.

Need for Sedimentation Prediction

The navigation improvement project study (USACE 2019) was undertaken to develop the effects of widening and deepening the Cape Fear River Navigation Channel. Environmental forcing data, such as measured water levels, annual and seasonal wind data, wave data, precipitation, river discharge and salinity, were used to model existing

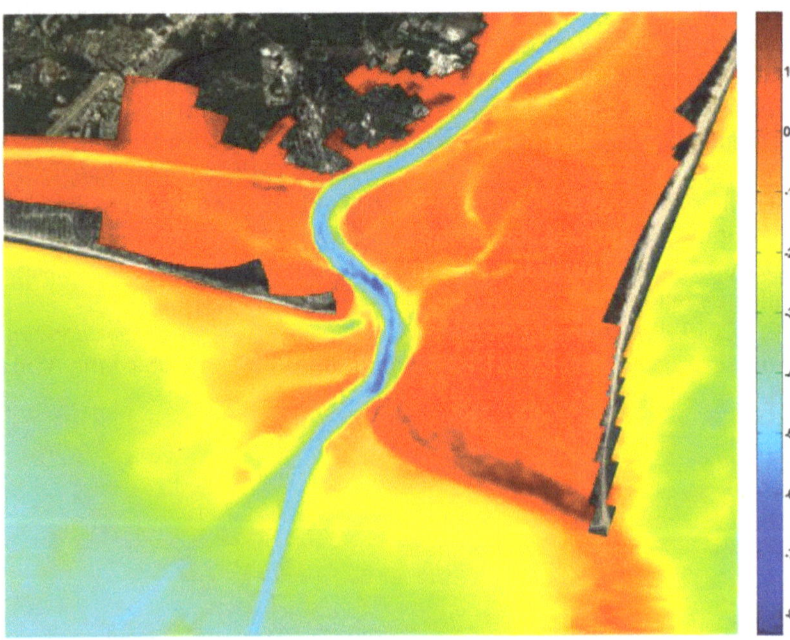

Figure B-13. Entrance channel morphology model bathymetry near the Cape Fear River entrance.
Source: USACE (2019).

conditions versus future conditions with modified channel geometry. This case study description will focus on the ocean entrance channel, Baldhead Shoals, Reaches 1 through 3, to describe the sediment prediction tools in the coastal, offshore setting. The study used a combination of GenCade for shoreline evolution and littoral transport patterns and Delft3D Waves for waves and sedimentation volumes within the navigation channel.

Sedimentation Prediction

The GenCade model (Frey et al. 2012) was applied to understand the historical longshore sediment transport and erosional patterns adjacent to the Cape Fear River Channel deepening area including both Bald Head Island and Oak Island/Caswell Beach (to the west and east of the entrance channel, respectively), and to evaluate the potential impact to oceanfront shorelines after the proposed channel deepening. The models for the shorelines were validated through the use of multiple reference shorelines and variation of sedimentation transport rates and offshore contour lines. In this use of GenCade, storm-induced cross-shore transport processes were excluded, which affected the validation results.

The validated GenCade was used to predict shoreline changes based on the future with project channel conditions caused by changes in wave patterns through the deepened channel, impacting the adjacent shores.

The Delft3D coupled Flow and Waves models (Deltares 2016) were used to develop sediment transport magnitudes along adjacent beaches and resulting sedimentation rates in the Cape Fear River entrance channels. To keep a reasonable model run time, a nested flow modeling approach was adopted for morphology modeling. Flow and wave grids were developed, and bathymetry was added from multiple recent surveys (the most recent survey for each area was used for the model). A schematized tide representation of average tidal fluctuations for each month through the year was developed to decrease model computing load. This schematized tide represents the average currents and sediment transport that occur during a spring-neap tide cycle but requires a slight correction factor. Offshore wave classes were developed through an optimum selection method (Mol 2007). Winds were used in the wave generation, but wind-generated currents were not considered. The modeling used a morphological time-step factor ranging from 0.5 to 17.6, depending on wave conditions. Spin-up time for the morphological modeling was 0.5 days. The models were validated on the basis of past measured channel shoaling patterns from three condition surveys in 2007, 2008, and 2010. Through the validation process, it was found that the shoaling volumes from the model were strongly dependent on the sediment grain size within each reach, and for model predictive use, grain sizes from 0.15 to 0.25 mm were adopted for specific reaches.

For prediction of sedimentation, the models were run for future without project conditions, and future with project conditions (a five-foot deepening in both riverine and offshore sections of the Federal channel). The shoaling volumes in the entrance channels were calculated after a 1-year simulation of sediment transport and morphology changes under annual average wave conditions. Tide and wave schematizations were the same as those used during model validation, and ranges of grain size of materials were considered. In the end, the averaged sedimentation increases caused by deepening within the various offshore channel reaches, derived from the Delft3D wave and flow modeling, varied from 4% to 9% annually, and for the offshore area as a whole, an 8% annual increase in sedimentation was predicted.

REFERENCES

Deltares. 2016. *Delft3D-FLOW user manual*. 3.15 ed. Delft, Netherlands: Deltares.

Frey, A. E., K. J. Connell, H. Hanson, M. Larson, et al. 2012. *GenCade version 1: Model theory and user's guide*. ERDC/CHL TR-12-25. Vicksburg, MS: US Army Engineer Research and Development Center.

Mol, A. C. S. 2007, *R&D Kustwaterbouw reductie golfrandvoorwaarden OPTI manual*. Research Rep. H4959.10. Delft, Netherlands: WL | Delft Hydraulics.

USACE (US Army Corps of Engineers). 2019. *Wilmington district, Wilmington Harbor, North Carolina navigation improvement project integrated section 203 study and environmental report*. Washington, DC: USACE.

LONG ISLAND, NEW YORK, SOUTH SHORE CHANNELS

Setting

The south Atlantic Ocean shoreline of Long Island (LI), New York, encompasses more than 100 mi of shoreline and within its length contains six shallow-draft US federal authorized navigation channels (Figure B-14). The primary use of these channels is for recreational watercraft; however, East Rockaway Inlet is used as a transit for commercial and residential fuel oil. The channels are also used to transit to bays and harbors of refuge for recreational water craft. The US Coast Guard is stationed at four of six inlets. The six channels are within inlets to back bays of various extents, and from west to east are: Rockaway (Jamaica Bay), East Rockaway, Jones, Fire Island, Moriches, and Shinnecock. The navigation channels beside the East Rockaway Inlet channel vary from 100 to 450 ft wide, with authorized depths from 10 to 12 ft MLLW (USACE 2022b). East Rockaway Inlet is 1,000 ft wide and has an authorized depth of 20 ft MLLW.

Need for Sedimentation Prediction

The navigation channels are subject to sedimentation from longshore transport, cross-shore transport from the offshore shelf, and from impacts of sediments moving from the ebb and flood shoals into the channel limits. A large volume of material from shoreline placement of offshore coastal sediments as part of multiple recently started federal coastal storm risk management projects will likely impact the channel maintenance operations in both frequency and volume. The ebb shoals adjacent to the navigation channels will serve in the future as additional sediment sources for shoreline material placement, both updrift and downdrift of the navigation channels.

Sedimentation Prediction

The USACE numerical 1D model GenCade (USACE 2022a) was used to predict changes in channel sedimentation rates. The model (Buonaiuto et al. 2021) incorporated an estimation of longshore transport rates. Sediment transport rates were examined from a multitude of historical regional sediment budgets. The shoreline was segmented into 14 subregions, and maximum, minimum, and average annual sediment transport rates were developed from historical rates. One to eight historical rates were available for each subregion, with 9 of the 14 subregions having four or more historical transport rate estimates. Winds, waves, astronomical tides, storm surge, and sea-level rise effects were examined. For the GenCade application, the 34-year wave record from 1980 to 2014 was

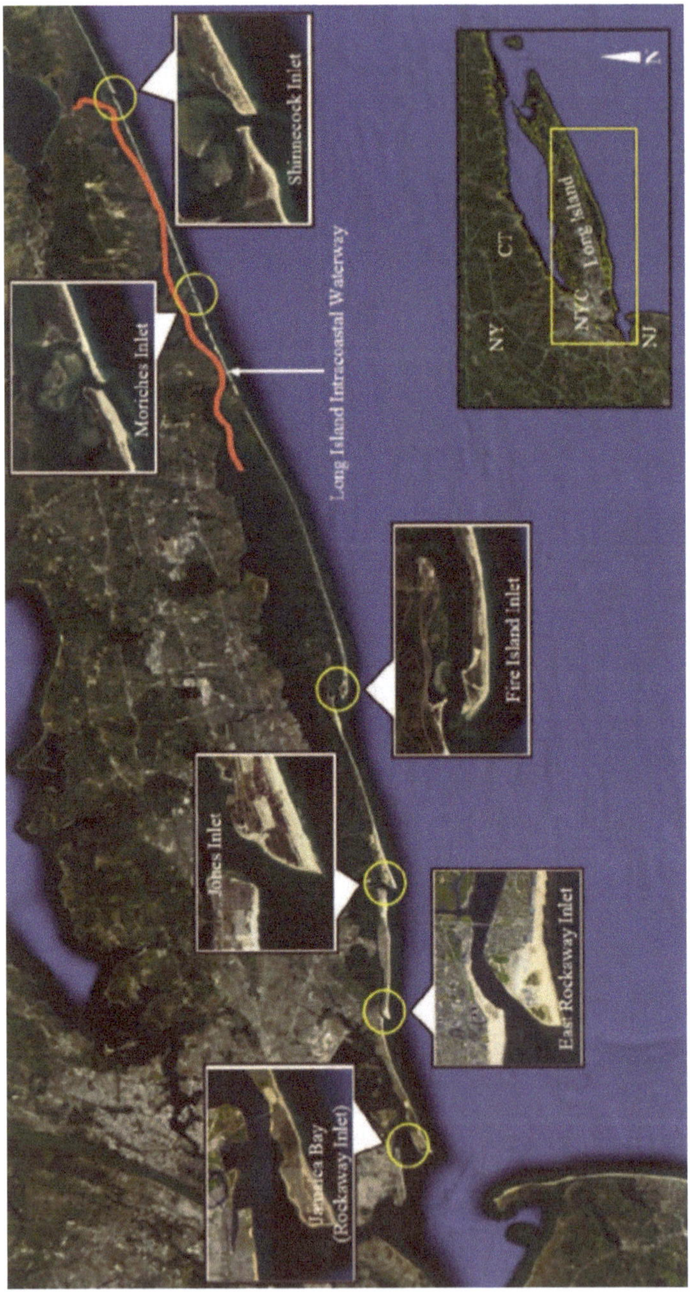

Figure B-14. Federal navigation channels on the south shore of Long Island. Source: Buonaiuto et al. (2021).

applied for both future without project conditions, and with project conditions. This wave period included a number of hurricanes and northeasters of record, including Hurricane Bob, the northeaster of 1992, and Hurricane Sandy. The future without project condition portrayed the status quo, in terms of quantity and frequency of navigation channel dredging and placement locations, and the continuation of two small storm risk management projects. The future with project condition incorporated the addition of coastal storm risk management projects at Rockaway (beachfill placement from offshore sources, additional groins), Long Beach (beachfill placement from offshore sources, rehabilitation, and addition of new groins), Fire Island to Montauk Point (beachfill placement in three locations from offshore sources, sand bypassing at the three east inlets, and removal of two groins). The future conditions also included more frequent and greater volume of sand removal from the authorized channels, and use of the ebb-shoal as beachfill material. A new, unstructured inlet within the Wilderness Area of Fire Island National Seashore was also included in the future conditions model.

The digital model setup included a regional 5.5 m offshore contour to represent the general shoreline orientation trend, an initial shoreline representing the starting condition of the system, the locations and properties of all engineering structures, initial inlet characteristics and equilibrium shoal volumes for use in the inlet reservoir model subroutine, quantities and locations of all dredging and beach nourishment events, and time series of wave forcing for the modeled period. In addition, the average berm height, depth of closure, and median grain size were assumed to be 2.5 m (approximately 8 ft), 8 m (approximately 25 ft), and 0.3 mm, respectively. The model was validated with measured shoreline data from 2016 to 2018 and compared to regional longshore transport studies. The model was also validated using LiDAR shoreline data collected in January 2020. The simulation was run from November 1, 2012 (immediately post-Hurricane Sandy) through January 1, 2020, with a 1 h time step. Default sand transport coefficients K1 and K2 were set to 0.27 and 0.3, respectively. The future without project and future with-project conditions simulation were then run.

Model results illustrated the connections between the region's shorelines, inlet navigation channels, and ebb shoals. Increased dredging of the authorized channels, and use of the ebb shoal material, contribute to larger sediment sinks for alongshore transport of coastal sediments, as more material is trapped in the channels and the shoals (instead of naturally by-passing those features), and may contribute to erosional shoreline areas directly west of the three eastern inlets. The modeling results demonstrated the need for continued monitoring of dredging and placement practices. The GenCade model provided a predictive tool for possible future conditions and led to a conclusion for the need for adaptive management of the region's inlet channels, ebb shoals, and shorelines.

REFERENCES

Buonaiuto, F. S., A. Heer, and S. Rice. 2021. *Regional sediment management: South shore of long island, GENCADE modeling summary*. Draft 2021. Vicksburg, MS: ERDC/CHL, US Army Engineer Research and Development Center.

USACE (US Army Corps of Engineers). 2022a. *GenCade, coastal inlets research program*. Vicksburg, MS: US Army Engineer Research and Development Center.

USACE. 2022b. *Controlling depth reports*. New York: USACE.

TENNESSEE–TOMBIGBEE WATERWAY ABERDEEN POOL

Setting

The Tennessee–Tombigbee (Tenn–Tom or TTW) Waterway is a 234 mi-long inland waterway providing a navigation connection between the Tennessee River (and thus the Cumberland, Ohio, and Mississippi Rivers) and the Gulf of Mexico via the Black Warrior-Tombigbee Waterway and Mobile Bay. It passes through Mississippi and Alabama, as shown in Figure B-15. The US Army Corps of Engineers began construction in December 1972 and completed the project in December 1984.

The project's Congressionally authorized purposes include navigation and wildlife conservation. It also provides extensive recreation and some water supply benefits to the extent that those uses do not adversely affect the authorized purpose benefits. It is not intended to provide flood damage

Figure B-15. Tenn–Tom Waterway.
Source: Courtesy of US Army Corps of Engineers.

reduction benefits; however, some residents of the Tombigbee River Valley assert that flood damage is noticeably reduced since the construction of the waterway.

Prior to waterway construction, the Tombigbee River was reliably navigable only to Columbus, MS, about 300 river miles upstream from Mobile Bay. Controlling depths were 1 ft or less during low-water seasons and about 4 ft during high-water season. The 20 mi reach between Columbus and Aberdeen, Mississippi, was navigable only during high–water seasons, but adverse currents made navigation extremely hazardous. Most navigation upstream of Columbus was limited to recreational vessels and log rafting (Mississippi SPC 1937).

The waterway consists of three distinct sections—River, Canal, and Divide Cut. The River portion extends upstream from Mile 217, where the Waterway connects to the Black Warrior River, to Mile 365 near Amory, Mississippi, generally following the course of the Tombigbee River. The Canal section starts at Mile 365 and departs from the Tombigbee River course to trend generally northward to Jamie Whitten (Bay Springs) Lock at Mile 412. The Divide Cut section connects the Canal section to the Tennessee River at Pickwick Lake near the Mississippi–Tennessee boundary.

The 149 mi-long River section lies within the Tombigbee River flood plain and generally follows the course of the river. A number of river meanders have been cut off, leaving 71 mi of meander loops that are still connected to the Waterway. Four lock and dam structures raise the water level to a total of 117 ft. The navigation channel has a bottom width of 300 ft and dredged depths of 9 or 12 ft plus 1 ft of allowable overdepth dredging. Numerous tributaries drain into the River section, bringing significant quantities of sediment.

The 46 mi-long Canal section is located near the eastern edge of the Tombigbee River floodplain and was formed by constructing a levee to serve as the western boundary of the section, whereas natural high ground serves as the eastern boundary. Five pools result in a chain-of-lakes configuration to provide navigable depths with a 300 ft-wide by 12 ft-deep channel. Inflow to the Canal section is limited to discharges from Whitten Lock and small tributaries on the eastern edge of the floodplain.

The Divide Cut section connects the separate river basins by an excavated cut through the basin divide and extends 39 mi from Bay Springs Lock to Pickwick Lake. The navigation channel has a bottom width of 280 ft and a depth of 12 ft during minimum (winter) pool on Pickwick Lake. Inflows to the section consist of minor local inflows and flow from Pickwick Lake to replace water released downstream at Whitten (Bay Springs) Lock.

The hydrology of the Tenn–Tom is unusual in that the waters of two basins are mixed by the connection at Bay Springs Lock. There, the waters of the Tennessee River, at least the volumes involved in lockages, are conveyed to the previously separate Tombigbee River valley. This aspect—the mixing of the waters—was one element in a vigorous debate over the Waterway's environmental effects prior to and during its construction and the title of a book on the project (Stine 1993).

Need for Sedimentation Prediction

Prior to construction of the Waterway, the Tombigbee River carried an estimated 239,000 t of sediment per year at Gainesville, Alabama (Underwood and Imsand 1985). Much of the watershed experienced severe land erosion and gullying (Mississippi SPC 1937). Watershed sedimentation issues for the completed waterway are much the same as preconstruction—sheet erosion and gullying during high runoff events. The tributaries of Town and Matubby Creeks experience severe channel incising and bank failures, both of which supply substantial quantities of sediment to the waterway. The lower Tombigbee, from Mile 279 seaward, also experiences bankline erosion. Bankhead et al. (2008) examined the bank erosion and could not separate Tenn–Tom Waterway effects, including vessel wakes, from natural causes (e.g., an increase in rainfall over the basin), and prewaterway impoundments (Bankhead et al. 2008).

Maintenance dredging quantities for the Waterway are about 820,000 cu yd per year, of which about 220,000 cu yd per year occurs in Aberdeen Pool (Sharp 2007).

Sedimentation Prediction

During the waterway design phase, the USACE Mobile District performed a field and desktop sedimentation analysis of the Tombigbee River reach that would become Aberdeen Pool. Flow-velocity measurements plus suspended sediment and bed samples were collected at three locations, one on each of two major tributaries and one on the Tombigbee River at the planned lock and dam site. The design memorandum (USACE 1979) provided analyses of historical data plus new data. It predicted postconstruction sediment delivery and, using a trapping efficiency of about 43% derived from similar lakes, estimated total sediment trapping and required channel maintenance dredging, shown in Table B-3.

A postconstruction desktop analyses of hydrographic surveys from 1985 to 2006 and a sediment budget analysis of Aberdeen Pool using regression on updated USGS data produced estimates for sediment delivery and total deposition, as shown in the third and fourth columns of Table B-3 (Sharp 2007).

Table B-3. Sediment Deposition Rates in Aberdeen Pool (1,000 t/year).

Category	Design (USACESAM 1979)	Sediment budget (Sharp 2007)	Hydrographic surveys 1985–2006 (Sharp 2007)	USACE records
Inflow supply	1,550	4,250	—	—
Total deposition	668	568–838	932	—
Dredging	461	—	—	305

The method consists of three steps and produced deposition estimates as shown:

1. Power-law regression equation for annual sediment transport using USGS data—838,000 t per year deposition.
2. Power-law regression equation for daily transport using USGS data—568,000 t per year deposition.
3. Sediment Impact Assessment Model (SIAM)—624,000 t per year deposition.

Sharp (2007) also analyzed hydrographic survey data from the Pool and calculated average overall deposition of 932,000 t per year, of which 305,000 t was removed by dredging, leaving 627,000 t per year net accumulation.

USACE records for 1985 to 2006 showed average annual maintenance dredging, as shown in the 4th column of Table B-3. It can be seen that the design document substantially underestimated sediment supply to and total deposition in the pool; however, dredging prediction was only about 50% higher than actual. The difference between total sediment supply and deposition (much greater than estimated) and dredging requirement (somewhat higher than estimated) is believed to be caused by vessel propwash and pressure wave resuspending sediment within the channel, which then deposited in the pool outside the channel.

Clifton (2011) used 2DH and 3D digital models meshes with modeling program EFDC to calculate sedimentation in Aberdeen Pool. The two models produced very similar water levels, sediment concentrations, and shear stresses. Bed changes were similar, differing only in the rates of change, with the 2D model eroding less and depositing more than the 3D model. A 4 s time step was employed for the 2D model, but the 3D version required smaller time steps for stability. No attempt was made to reproduce historical conditions, as reported in Table B-3.

REFERENCES

Bankhead, N., A. Simon, and D. Klimetz. 2008. *Analysis of streambank erosion along the lower Tombigbee River, Alabama*. Research Rep. No. 62. Oxford, MS: USDA-ARS National Sedimentation Laboratory.

Clifton, N. D. 2011. "Numerical analysis of Aberdeen pool sedimentation." Master's thesis, Mississippi State University, Dept. of Civil and Environmental Engineering.

Mississippi SPC (State Planning Commission). 1937. *Report on the Tombigbee River drainage basin*. Jackson, MS: Mississippi SPC and Works Progress Administration.

Sharp, J. A. 2007. "Sediment budget template applied to Aberdeen pool." Master's thesis, Mississippi State University, Dept. of Civil and Environmental Engineering.

Stine, J. K. 1993. *Mixing the waters: Environment, politics, and the building of the Tennessee-Tombigbee Waterway*. Akron, OH: University of Akron Press.

Underwood, K. D., and F. D. Imsand. 1985. "Hydrology, hydraulic, and sediment considerations of the Tennessee-Tombigbee Waterway." *Environ. Geol. Water Sci.* 7 (1–2): 69–90.

USACE (US Army Corps of Engineers). 1979. *Tennessee-Tombigbee Waterway Alabama and Mississippi, Aberdeen lock and dam*. Design Memorandum No. 17: Sedimentation Program. Mobile, AL: USACE, Mobile District.

ATCHAFALAYA BAY CHANNEL

Setting

Atchafalaya Bay, Louisiana, is part of an extensive interconnecting system of tidal bays, which include East Cote Blanche, West Cote Blanche, and Vermillion Bays. The system opens to the Gulf of Mexico along the 27 mi southern boundary of Atchafalaya Bay and also through the narrow Southwest Pass inlet of Vermillion Bay. Tidal amplitudes are typically less than 2 ft, and therefore, bay hydrodynamics and sediment transport are principally driven by river flow, wind, and waves.

The Atchafalaya River captures about 30% of the combined flow of the Mississippi River and Red River at the latitude of the Old River Diversion Structure, and carries to Atchafalaya Bay an average of 94 million tons of sediment in suspension every year. That sediment load progressively filled in the Atchafalaya basin floodway between its levee systems and then began depositing rapidly in Atchafalaya Bay (Figure B-16).

Need for Sedimentation Predictions

The Atchafalaya Bay navigational channel, part of the Atchafalaya River and Bayous Chene, Boeuf, and Black navigation channel, bisects the Bay, extending from the mouth of the Lower Atchafalaya River through to west of Point Au Fer and thence to the offshore Bar Channel. Historically, in a low river discharge year, little or no dredging was normally required in the bay channel. However, the high discharge years of 1973 to 1975 necessitated dredging of more than 180 million yd3, and periodic dredging is now required.

That dredging requirement and the importance of land growth for Louisiana led the Corps of Engineers to evaluate how the delta and dredging need would evolve over the following decades.

Sedimentation Prediction

The prediction approach used field studies plus desktop, physical, and numerical modeling in a tiered approach (McAnally et al. 1991). Several methods applied separately allowed cross-comparison and error-checking of projections. The methods included

- Field measurements of bathymetry, water level, velocity, salinity, and sediment (e.g., Teeter and Pankow 1989).
- Desktop extrapolation of observed bathymetric changes into the future, adjusted for apparent subsidence (Letter 1982).

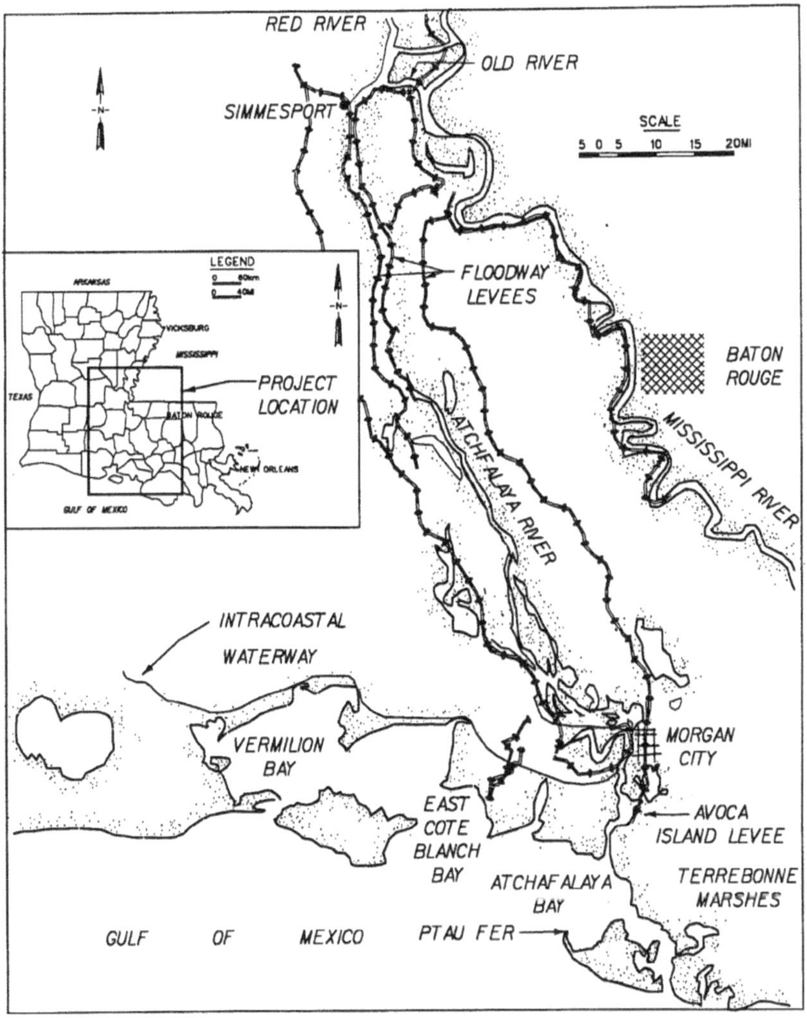

Figure B-16. Atchafalaya Bay site map.
Source: Donnell and Letter (1992).

- Desktop generic future delta growth by constructing an analogy between behavior of the Atchafalaya delta and other deltas in similar environments (Wells et al. 1984).
- Jet model analytical solution of a sediment-laden jet discharging into a quiescent bay (Wang 1985).
- Quasi-two-dimensional numerical modeling of hydrodynamics and sedimentation processes for the river flowing into a quiescent bay using HAD-1 (Thomas et al. 1988).

- One-dimensional network modeling of deltaic channels using SOCHMJ (Thomas et al. 1988).
- Physical modeling of flow patterns in the bay and delta.
- Two-dimensional (2DH) numerical modeling of hydrodynamics and sedimentation processes forced by river flow, tides, Gulf levels, storm surges, wind-induced currents, wind waves, salinity, and subsidence using Tabs-2 (Donnell et al. 1991).

Quasi-2D modeling performed continuous simulations for 50 years delta growth with a computational mesh consisting of 20 lateral segments (1 to 5 mi apart), each divided into seven longitudinal strips.

The 2DH modeling (Donnell et al. 1991) used an event modeling approach to construct "morphological" time steps, although they did not use that term. Events were assembled from the following conditions:

- Repeating mean diurnal tide,
- Four river discharges,
- Two gulf levels,
- Cohesive and noncohesive sediment inflow,
- Four inflowing sediment concentrations,
- Three wave regimes, and
- Three salinity regimes.

Each event was simulated for a year with a 30 min hydrodynamic time step and a 60 min sediment model time step. Sedimentation spin up consisted of a 1-year simulation with mean boundary and hydrodynamic conditions. Multiple sensitivity tests examined the effect of model parameters, mesh resolution, time step, and boundary conditions to choose appropriate values.

Model validation consisted of simulating years 1967 to 1977 with five events combined via joint probabilities. The results were consistent with observed delta extent and dredging volumes and with comparable 10-year sediment accumulation generated by the desktop and quasi-2D results.

Fifty-year sedimentation projections consisted of the following steps:

1. Start with 1980 bathymetry and spun-up flow and sediment concentrations.
2. Simulate seven events.
3. Combine Step 2 events by joint probabilities to extrapolate 15 years bathymetric change and dredging quantities.
4. Simulate seven new events starting with bathymetry from prior step.
5. Combine Step 4 events via joint probabilities to extrapolate 15 or 20 years bathymetric change and dredging quantities.
6. Repeat Steps 4 and 5 until 50 years are simulated.

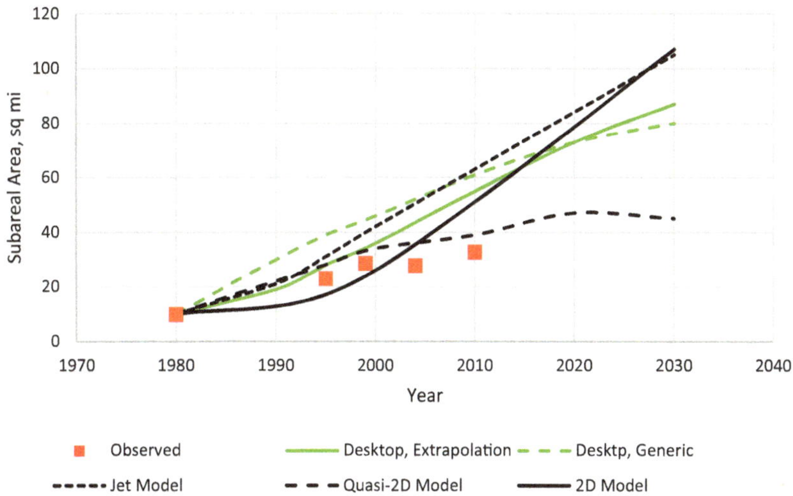

Figure B-17. Atchafalaya Bay Delta Subaerial Extent.
Source: Model data from Donnell and Letter (1992) and observed data from Rosen and Xu (2013).

Model plan testing consisted of five plans consisting of combinations from an existing levee extension, outlet flow control, and dredge material placement to create levees on both sides of the navigation channel. Fifty-year simulations, as described previously, projected delta growth and channel dredging requirements for each plan. Projected dredging volumes, expressed as a ratio of plan volume to existing dredged volume (rounded to reflect uncertainty), showed that none had a large effect on dredging need. However, all plans showed substantial reduction in dredging requirements after 30 years because of increased channelization of flow through the growing delta (Donnell and Letter 1992).

Delta growth predictions from the several prediction approaches are compared to observed subaerial delta extent in Figure B-17. It can be seen that the predictions of two of the approaches have been higher than the observed data so far. The desktop extrapolation, quasi-2D modeling, and 2DH modeling predictions were all close to observed up to year 2000, but only the quasi-2D modeling predicted the slowdown in delta growth from 2000 to 2010.

REFERENCES

Donnell, B. P., and J. V. Letter Jr. 1992. *The Atchafalaya River Delta: Report 12, two-dimensional modeling of alternative plans and impacts on the Atchafalaya Bay and Terrebonne Marshes*. Tech. Rep. HL-82-15. Vicksburg, MS: US Army Engineer Waterways Experiment Station.

Donnell, B. P., J. V. Letter Jr., and A. M. Teeter. 1991. *The Atchafalaya River Delta: Report 11, two-dimensional modeling*. Tech. Rep. HL-82-15. Vicksburg, MS: US Army Engineer Waterways Experiment Station.

Letter, J. V., Jr. 1982. *The Atchafalaya River Delta: Report 3, extrapolation of delta growth*. Tech. Rep. HL-82-15. Vicksburg, MS: US Army Engineer Waterways Experiment Station.

McAnally, W. H., S. B. Heltzel, and B. P. Donnell. 1991. *The Atchafalaya River Delta: Report 1, a plan for predicting the evolution of Atchafalaya Bay, Louisiana*. Tech. Rep. HL-82-15. Vicksburg, MS: US Army Engineer Waterways Experiment Station.

Rosen, T., and Y. J. Xu. 2013. "Recent decadal growth of the Atchafalaya River Delta complex: Effects of variable riverine sediment input and vegetation succession." *Geomorphology* 194 (July): 108–120.

Teeter, A. M., and W. Pankow. 1989. *The Atchafalaya River Delta: Report 2, field data; section 2: Settling characteristics of bay sediments*. Tech. Rep. HL-82-15. Vicksburg, MS: US Army Engineer Waterways Experiment Station.

Thomas, W. A., R. E. Heath, J. P. Stewart, and D. G. Clark. 1988. *The Atchafalaya River Delta: Report 5, the Atchafalaya River Delta quasi-two-dimensional model of delta growth and impacts on river stages*. Tech. Rep. HL-82-15. Vicksburg, MS: US Army Engineer Waterways Experiment Station.

Wang, F. C. 1985. *The Atchafalaya River Delta: Report 7, analytical analysis of the development of the Atchafalaya River Delta*. Tech. Rep. HL-82-15. Vicksburg, MS: US Army Engineer Waterways Experiment Station.

Wells, J. T., S. J. Chinburg, and J. M. Coleman. 1984. *The Atchafalaya River Delta: Report 4, generic analysis of delta development*. Tech. Rep. HL-82-15. Vicksburg, MS: US Army Engineer Waterways Experiment Station.

DOGTOOTH BEND REACH, MISSISSIPPI RIVER

Setting

In the 1980s, the reach of the Mississippi River from River Mile 39.6 to River Mile 20.2 (river miles are above the confluence of the Mississippi and Ohio Rivers at Cairo, Illinois) had several problems, including inadequate navigation widths in two bends, narrow or nonexistent navigation channels in the crossings downstream of the bends, and significant maintenance dredging to keep the navigation channel open. This reach of the Mississippi River is referred to as Dogtooth Bend Reach.

Need for Sedimentation Prediction

As mentioned previously, some of the major issues in the Dogtooth Bend Reach were the crossings/bends. Figure B-18 shows a portion of the prototype channel configuration from April 1983. The crossing in the vicinity of range R-47 has a bar above elevation +10 ft (referenced to the Low Water Reference Plane, LWRP) in the middle of the crossing. The crossing from range R-55 to R-57 is very narrow for navigation. The point bar on the left side of the channel (looking downstream) encroaches on the navigation channel on the right bankline from about range R-59 to R-64.

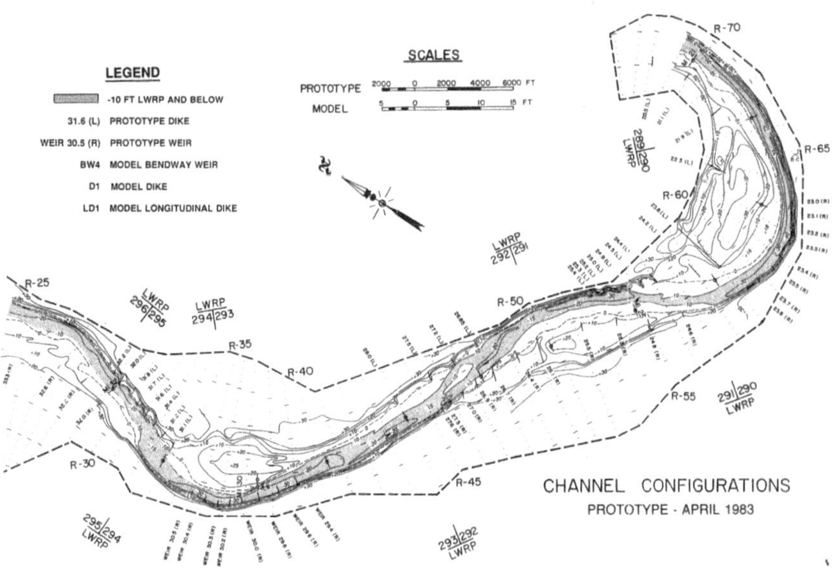

Figure B-18. 1983 Prototype Channel Configuration, Dogtooth Bend, Mississippi River.
Source: Courtesy of USACE.

All three of these problem areas would require maintenance dredging to obtain the required navigation channel dimensions of 9 ft deep (below the LWRP) and 300 ft wide.

Sedimentation Prediction

To address these and other issues, a physical movable-bed model was constructed in the Inland Waterways Facility of the Coastal and Hydraulics Laboratory (CHL) at the US Army Engineering and Research Center (ERDC) in Vicksburg, Mississippi. Figure B-19 shows a view of the facility. Conducting a model study in this facility has advantages in that the inflow pumps, outflow controls, model controls, and data collection are installed and readily available. This significantly reduces the cost of construction and operation of such a model study.

The movable-bed model used in the study used crushed coal as the bed material. The coal had a specific gravity of 1.30 and a mean grain size of 4 mm. Crushed limestone was used to model the banklines and river training structures (dikes, weirs, and revetments). The model was constructed to a horizontal scale of 1:400 and a vertical scale of 1:100, producing a distortion ratio of 4. The model reproduced the portion of the Mississippi River from River Mile 39.2 to 20.2 including portions of the overbank area up to the Mainline levee.

Figure B-19. Inland Waterways Research Facility, Coastal and Hydraulics Lab, ERDC.
Source: Courtesy of USACE.
Note: This is not a photograph of the Dogtooth Bend Model.

The initial testing conducted on the Dogtooth Bend Reach model consisted of numerous tests composed of a variety of longitudinal spur dikes, stub dikes, detached dikes, and/or curved dikes. All these tests conformed to the minimum contraction width of 1,500 ft for this portion of the Mississippi River, that is that no structures can be built to confine the river narrower. None of these initial tests produced the desired results for the study.

In an unrelated general research study, CHL engineers had investigated using a series of upstream angled weirs (subsequently referred to as *bendway weirs*) submerged in the outside (deepest portion of the navigation channel) of a bend. The weirs had a crest elevation low enough to allow vessels to pass unimpeded over them. On the basis of general research, it appeared that the weirs disrupted the secondary currents and helicoidal flow in the bends and redirected the flow passing over the weirs toward the opposite point bar. It was decided to investigate this new concept of river training in the Dogtooth Bend study.

The bendway weirs were installed on the right bankline upstream of range R-40 (not shown on Figure B-18) and on the right bend bankline from range R-57 to R-65. The bendway weirs were level-crested at elevation −15 ft LWRP, generally spaced 1,500 ft apart, and angled upstream 30 degrees. The maximum bendway weir height was based on a tow operating in low water and drafting 9 ft with a factor of safety added for ice buildup on the bottom of barges during cold weather.

The channel configuration at the end of the bendway weir tests is shown on Figure B-20. The results show that the weirs installed upstream of range R-40 improved the crossing and removed the middle bar at range R-47. Also, the channel improvement in this area widened and deepened the crossing at range R-55 to R-57. Finally, the bendway weirs installed between range R-57 and R-65 widened the navigation channel in this area an average of 200 ft. In addition, 15 to 30 ft of sediment was deposited on the outside of the bend, which added stability to the bank.

In the summary of the bendway weir model tests, it was stated that the weir benefits included a wider and deeper navigation channel at the lower bend, especially at the apex; the downstream bend was stabilized because of sediment being deposited on the slope toe; the navigation downstream of the downstream bend was deeper and wider; the crossings upstream at ranges R-40 and R-47 were deeper and better aligned; surface water velocities were uniform across all bendways; and the flow patterns in the bends were parallel with the banks and did not concentrate along the outside of bends (Derrick et al. 1994). Figure B-21 is a photograph of a portion of one of the bendway weir fields in the model.

The bendway weir plan developed in the coal-bed movable model was installed, with some modifications in the prototype in the late 1990s. Besides all observed benefits listed previously, the reduction in dredging

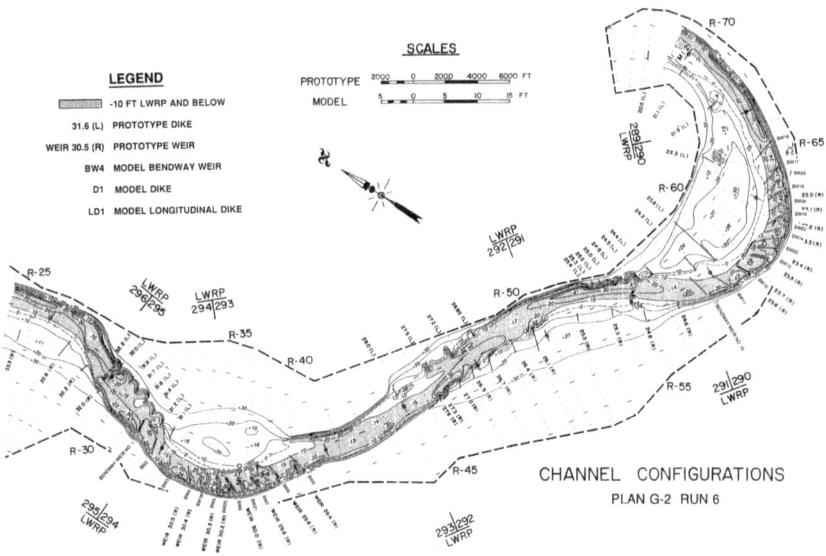

Figure B-20. Channel Configuration at the End of Bendway Weir Testing, Dogtooth Bend, Mississippi River.
Source: Courtesy of USACE.

Figure B-21. Bendway Weir Layout at Dogtooth Bend, Mississippi River.
Source: Courtesy of USACE.

in the two bends described above was almost unbelievable. Prior to installation of the bendway weirs in those bends, the St. Louis District of the Corps of Engineers dredged almost 1.8 million yd3 of sediment per year. That dredged amount was reduced to 56,000 yd3 per year following the installation of the weirs. That was a cost savings to the district of $400,000 for dredging alone (Fatherree 2004). Because of the success of the bendway weir concept, weirs have been installed in many other reaches of the Mississippi River with similar success.

REFERENCES

Derrick, D. L., T. J. Pokrefke Jr., M. B. Boyd, J. P. Crutchfield, et al. 1994. *Design and development of bendway weirs for the Dogtooth Bend Reach, Mississippi River*. Tech. Rep. HL-94-10. Washington, DC: USACE Waterways Experiment Station.

Fatherree, B. H. 2004. *The first 75 years: History of hydraulics engineering at the Waterways Experiment Station*. Vicksburg, MS: USACE Engineer Research and Development Center.

APPENDIX C
DECISION TREE TOOL SELECTION

Chapter 6, "Best Practices," describes considerations for choosing prediction methods listed in Chapter 6, "Approach and Tools Selection" section. This appendix provides a decision tree structure for the process. The following steps, illustrated in Figures C-1 through C-3, are keyed to the step and symbol numbers in the figures.

Strengths and limitations of each tool, as described in Chapters 5 and 6, should be considered along with the criteria listed here. In this simplified decision tree, the possible need for multiple models, as described in Chapter 6, are implied by using the plural term.

Step 1: Define the Situation

Follow process as described in the "Problem Definition" and "Approach and Tools Selection" sections of Chapter 6.

Step 2: Evaluate the Situation

2.1. Processes
Are hydrodynamics and sediment transport complicated by rapidly varying geometry—widths, depths, or alignment? Are there structures that strongly influence sediment transport? Is sedimentation around structures an issue? Does lateral distribution of sediment transport and deposition/erosion affect the problem and possible solutions? Should the results be predicted for decades

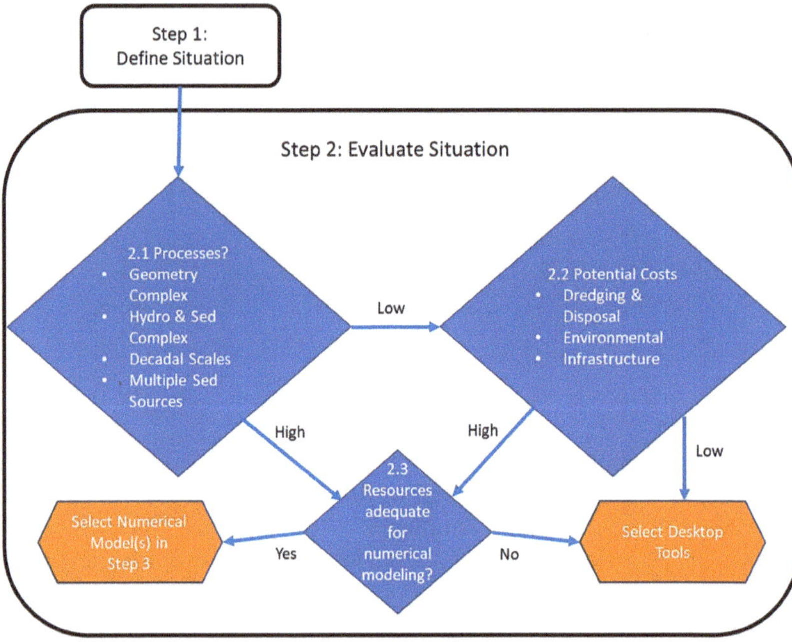

Figure C-1. Decision tree Steps 1 and 2: Define and evaluate situation.

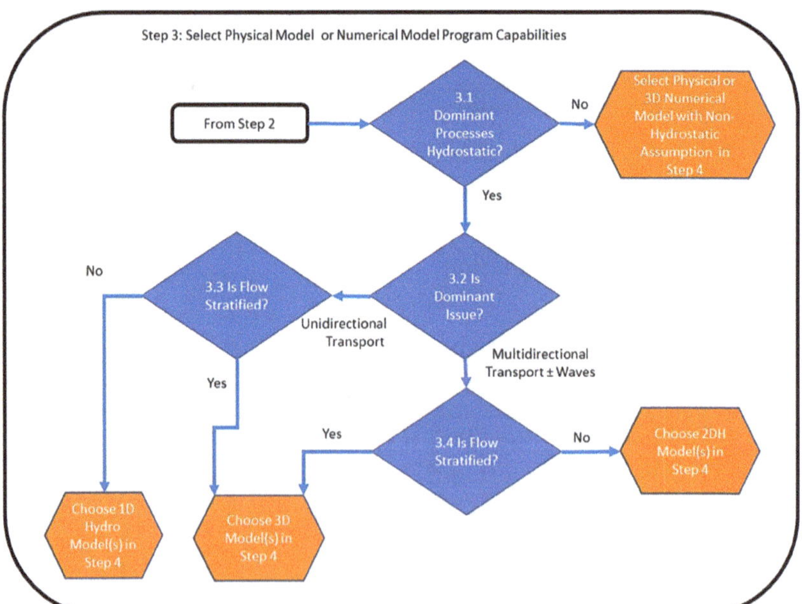

Figure C-2. Decision tree Step 3: Select model capabilities.

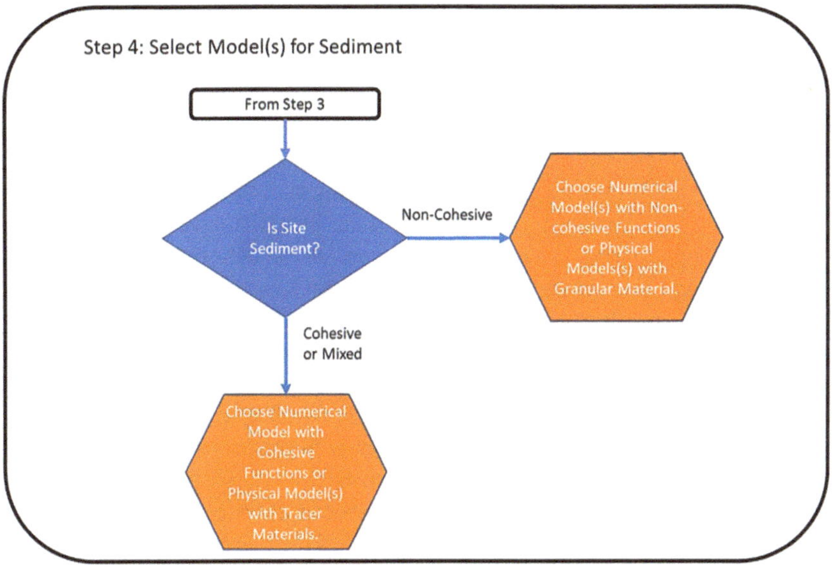

Figure C-3. Decision tree Step 4: Model program selection.

or longer? Are there multiple sediment sources to the problem area? If answers to these questions are

- No (low complexity), go to 2.2.
- Yes (high complexity), go to 2.3.

2.2. Potential Costs

Cost considerations here are qualitative and not quantified estimates. They are also context-dependent, in that "high" and "low" costs for a federal agency may be much different from those for a local agency or a nongovernmental organization.

Are either the sedimentation costs high or expected to be high (expensive dredging, valuable infrastructure threatened, etc.) as defined in Step 1? Are potential solutions defined in Step 1 expected to be of high cost to build or maintain? Do the problem and potential solutions have a high likelihood of affecting environmental quality? If answers to these questions are

- Low, select desktop tools.
- High, go to 2.3.

2.3. Resources Adequate for Modeling

Are the available personnel, equipment, and allowed time and cost for the evaluation sufficient for numerical modeling?

- If no, select desktop tools.
- If yes, go to Step 3.

Step 3. Select a Physical Model or Numerical Model Program Capabilities

3.1 Can dominant hydrodynamic processes at site be treated as hydrostatic?

This decision requires the judgement of an experienced hydrodynamic engineer. Many geophysical flow situations in rivers, lakes, estuaries, and coastal areas can be satisfactorily addressed by using the hydrostatic assumption, in which it is assumed that vertical accelerations are small and pressure in the vertical can be approximated by the hydrostatic equation. The assumption has been widely and successfully used; however, it is not applicable in flow over and through structures such as dams or very near structures such as dikes and bridge piers. Fully defining such flows and resulting sediment transport requires that a nonhydrostatic model be employed. If the answer to this question is

- No, select 3D nonhydrostatic model in Step 4.
- Yes, go to 3.2.

3.2 Is the dominant sediment transport

- Unidirectional, as in a river or canal without significant lateral variation or problems of interest? Go to 3.3.
- Multidirectional, as in a wide waterway with significant lateral variation or tides, which may include wave action? Go to 3.4.

3.3 Is sediment transport substantially affected by flow that is stratified by salinity, temperature, or suspended sediment concentration?

- If no, choose 1D model(s) in Step 4.
- If yes, choose 3D model(s) in Step 4.

3.4 This is the same question as 3.3. Is sediment transport substantially affected by flow that is stratified by salinity, temperature, or suspended sediment concentration?

- No, choose 2D model(s) in Step 4.
- Yes, choose 3D model(s) in Step 4.

Step 4. Select Models for Sediment

If site bed and transport issues concern sediment that is

- Noncohesive, choose numerical model(s) with suspended transport and noncohesive suspended load, bedload, and bed exchange functions suitable for hydrodynamic conditions or physical model(s) with granular material.
- Cohesive or mixed, choose numerical model(s) with suspended transport and cohesive bed exchange functions suitable for hydrodynamic conditions or physical model(s) with tracer material.

APPENDIX D
ACRONYMS AND ABBREVIATIONS

This list provides the full names for acronyms and abbreviations, except for numerical model programs. Some numerical program full names are descriptive of their function (e.g., HEC-RAS), some are words only (e.g., Tabs), and others use creative wording to generate memorable acronyms (e.g., SOCHMJ). Here, a descriptive definition of numerical model program names is provided. Details on these programs can be found in references cited in Table 5-3. Some programs are part of suites or integrated collections of models and utility programs and are labeled as such.

1D	One-dimensional
2D	Two-dimensional
3D	Three-dimensional
AAPA	American Association of Port Authorities
ADCIRC	Ocean Circulation Hydrodynamic Model Program
AdH	2D Hydrodynamic and Sediment Model Program
AdH3	3D Hydrodynamic and Sediment Model Program
AOC	Atlantic Ocean Channel
APHA	American Public Health Association
ARS	Agricultural Research Station
ASTM	American Society for Testing and Materials
BASINS	Better Assessment Science Integrating Point and Nonpoint Sources
BNWL	Battelle Pacific Northwest Labs
BSS	Brier Skill Score
BSTEM	Bank Erosion Tool
CCHE2D	2D Hydrodynamic and Sediment Model Program
CCHE3D	3D Hydrodynamic and Sediment Model Program

CDIP	Coastal Data Information Program
CDS	Center for Dredging Studies
CERC	Coastal Engineering Research Center
CFR	Cape Fear River
CH3D	3D Hydrodynamic and Sediment Model Program
CHETN	Coastal and Hydraulics Engineering Technical Note
CHL	Coastal and Hydraulics Laboratory
CMS	2D and 3D Hydrodynamic and Sediment Model Program Suite
COPRI	Coasts, Oceans, Ports, and Rivers Institute
CSAT	Channel Shoaling Analysis Tool
CSTORM	Computer Program suite for modeling coastal storms
CTH	Committee on Tidal Hydraulics
DEM	Digital elevation model
DHI	Danish Hydraulic Institute
DOER	Dredging Operations and Environmental Research
DRP	Dredging Research Program
DWS	Deep Water Site
ECIP	Expansion Channel Improvement Project
ECOM	3D Hydrodynamic and Sediment Model Program
EFDC	3D Hydrodynamic and Sediment Model Program
EFDC_Explorer	GUI for EFDC produced by Dynamic Solutions International
EPA	Environmental Protection Agency
EPS	Extracellular polymeric substances
ERDC	Engineering Research and Development Center
ETM	Estuarine Turbidity Maxima
EWN	Engineering with Nature
EWRI	Environmental and Water Resources Institute
FISP	Federal Interagency Sedimentation Project
FLUENT	3D Hydrodynamic Model Program
FNC	Federal Navigation Channel
FTU	Formazin Turbidity Units
FUNWAVE-TVD	Wave Model Program
FVCOM	3D Hydrodynamic and Sediment Model Program
FWOP	Future without project
FWP	Future with project
GENCADE	1D Wave Model Program
GIS	Geographical information system
GLUE	Generalized Likelihood Uncertainty Estimation
GOM	Gulf of Mexico
GPS	Global positioning system
GRR	General Reevaluation Report
GSMM	Georgia Stormwater Management Manual

GUI	Graphical user interface
HEC-RAS	Hydrologic Engineering Center's River Analysis Program
HOP	Head of Passes
HRS	Hydraulic Research Station
HSC	Houston Ship Channel
IADC	International Association of Dredging Companies
IHO	International Hydrographic Organization
IIHR	Iowa Institute of Hydraulic Research
ISI	Institute for Sustainable Infrastructure
ISO	International Organization for Standardization
JTU	Jackson Turbidity Units
LCR	Lower Columbia River
LIDAR	Light Detection and Ranging System for Elevation Measurement
LPR	Lower Passaic River
LWRP	Low Water Reference Plane
MCR	Mouth of Columbia River
MIKE	2D and 3D Hydrodynamic and Sediment Model Program Suite
MLLW	Mean low lower water
MLW	Mean low water
MRG	Mississippi River Geomorphology
MRGO	Mississippi River Gulf Outlet
MRSC	Mississippi River Ship Channel
NAP	National Academies Press
NHER	Norfolk Harbor Entrance Reach
NNC	Newport News Channel
NOAA	National Oceanic and Atmospheric Administration
NRCS	Natural Resources Conservation Service
NTU	Nephelometric Turbidity Unit
NYNJ	New York/New Jersey
NYNJH	New York/New Jersey Harbor
OpenFOAM	3D Hydrodynamic Model Program
PCB	Polychlorinated biphenyl
PDF	Probability distribution function
PIANC	World Association for Waterborne Transport Infrastructure
PIT	Pathway Interactive visualization Tool
PTM	Particle Tracking Model Program
PWOP	Present without project
PWP	Present with project
RMSE	Root mean square error
RSM	Regional sediment management

SBAS	Sediment budget analysis system
SCOTUS	Supreme Court of the United States
SEDA	Sediment erosion and deposition assessment
SEDLIB	Sediment Transport Library program
SHOALS	Airborne LiDAR Bathymetry system
SIAM	1D Sediment Budget Tool
SMART	USACE Research Program
SMS	GUI for 1D, 2D, 3D Hydrodynamic and Sediment Model Program Suite
SOW	Statement of Work
SPIT	Sediment Pathway Interactive visualization Tool
SRH	2D Hydrodynamic and Sediment Model Program
SSC	Suspended sediment concentration (measurement produced by ASTM 2000 standard)
STOWA	Foundation for Applied Water Research
SWAN	2D Wave Model Program
SWP	Southwest Pass
SWWRP	System-Wide Water Resources Research Program
Tabs	2D Hydrodynamic and Sediment Model Program Suite
TSC	Thimble Shoals Channel
TSS	Total suspended solids (measurement produced by APHA 1995 standard)
TTW	Tennessee–Tombigbee Waterway
TVA	Tennessee Valley Authority
UMR	Upper Mississippi River
UNESCO	United Nations Educational, Scientific, and Cultural Organization
USACE	United States Army Corps of Engineers
USACENAN	United States Army Corps of Engineers, New York District
USACESAM	United States Army Corps of Engineers, Mobile District
USACESWG	United States Army Corps of Engineers, Galveston District
USDA	United States Department of Agriculture
USGS	United States Geological Survey
WEDA	Western Dredging Association
WES	Waterways Experiment Station
WHNIP	Wilmington Harbor Navigation Improvement Project
WHOI	Woods Hole Oceanographic Institution

INDEX

FSC
www.fsc.org
MIX
Papier | Fördert gute Waldnutzung
FSC® C083411

Kolibri 360 GmbH
Ferdinand-Jühlke-Straße 7
99095 Erfurt, Deutschland
produktsicherheit@kolibri360.de